KT-165-369

PRINCIPLES

—OF—

HYDROGEOLOGY

SECOND EDITION

PRINCIPLES
—OF—
HYDROGEOLOGY
SECOND EDITION

by

Paul F. Hudak

LEWIS PUBLISHERS

Boca Raton London New York Washington, D.C.

Cover Photo by John R. Kuiper
(spring-fed farm pond in southeastern Cochabamba, Bolivia)

Library of Congress Cataloging-in-Publication Data

Hudak, Paul F.
   Principles of hydrogeology / by Paul F. Hudak. — 2nd ed.
         p.   cm.
      Includes bibliographical references and index. (p.).
      ISBN 1-56670-500-2 (alk. paper)
      1. Hydrogeology.   I. Title.
   GB1003.2.H83 1999
   551.49—dc21   99-040090
                                                                                              CIP

This book contains information obtained from authentic and highly regarded sources. Reprinted material is quoted with permission, and sources are indicated. A wide variety of references are listed. Reasonable efforts have been made to publish reliable data and information, but the author and the publisher cannot assume responsibility for the validity of all materials or for the consequences of their use.

Neither this book nor any part may be reproduced or transmitted in any form or by any means, electronic or mechanical, including photocopying, microfilming, and recording, or by any information storage or retrieval system, without prior permission in writing from the publisher.

The consent of CRC Press LLC does not extend to copying for general distribution, for promotion, for creating new works, or for resale. Specific permission must be obtained in writing from CRC Press LLC for such copying.

Direct all inquiries to CRC Press LLC, 2000 N.W. Corporate Blvd., Boca Raton, Florida 33431.

Trademark Notice: Product or corporate names may be trademarks or registered trademarks, and are used only for identification and explanation, without intent to infringe.

© 2000 by CRC Press LLC
Lewis Publishers is an imprint of CRC Press LLC

No claim to original U.S. Government works
International Standard Book Number 1-56670-500-2
Library of Congress Card Number 99-040090
Printed in the United States of America   2 3 4 5 6 7 8 9 0
Printed on acid-free paper

00   09929

PREFACE

The objective of this book is to introduce key concepts of hydrogeology in a concise, yet informative manner. Its intended audience is college students with no background in hydrogeology. The text assumes that readers have some familiarity with introductory geology and algebra. Practitioners and environmental regulatory officials may also find the book useful, as a reference source, while working on groundwater problems.

ABOUT THE AUTHOR

Paul F. Hudak is a faculty member in the Department of Geography and Environmental Science Program at the University of North Texas. He received a B.S. in Geology from Allegheny College, an M.S. in Geology from Wright State University, and a Ph.D. in Geography from the University of California, Santa Barbara. Dr. Hudak has practiced hydrogeology in Pennsylvania, Ohio, California, and Texas. His primary research interests include hydrogeology, environmental monitoring and restoration, and geologic hazards.

For Sheri

TABLE OF CONTENTS

CHAPTER 1

Groundwater and the Hydrologic Cycle

Hydrogeology is the study of underground water. Underground water occurs in two different zones (Figure 1.1). The vadose zone, which immediately underlies the land surface, contains both water and air. Beneath the vadose zone lies the saturated zone, in which all of the interconnected openings are full of water. Water stored in the saturated zone is called groundwater.

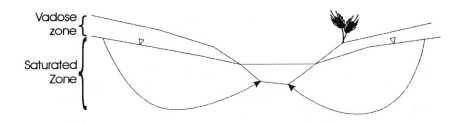

Figure 1.1. Distribution of underground water; inverted triangles mark water table.

Worldwide, groundwater is an important source of water for humans. Groundwater accounts for approximately 42% of the public and domestic water supplies in the United States (Solley and others, 1998). However, many cities, communities, and rural households rely entirely upon groundwater. San Antonio (TX), El Paso (TX), Albuquerque (NM), and Dayton (OH) are examples of large cities that depend almost completely upon groundwater. Other cities use groundwater to augment surface water sources such as lakes and rivers. For communities that rely entirely upon surface water, groundwater may contribute to the water in rivers and lakes.

Groundwater is but one component of the hydrologic cycle: the continuous circulation of water near Earth's surface (Figure 1.2). This cycle includes both reservoirs in which water is stored, and processes that transfer water between reservoirs. The major reservoirs of the hydrologic cycle are listed in Table 1.1. Groundwater accounts for the vast majority of freshwater available to humans.

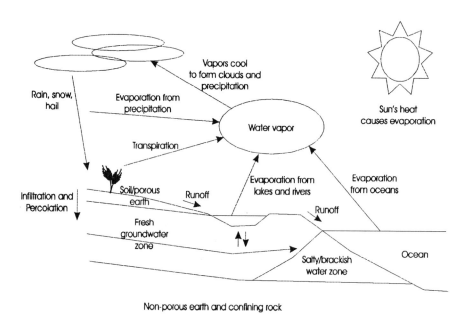

Figure 1.2. Schematic view of hydrologic cycle.

Table 1.1. Distribution of Earth's water (modified from Nace, 1967).

Reservoir	Amount (%)
oceans	97.2
ice caps and glaciers	2.14
groundwater	0.62
freshwater lakes	0.009
saline lakes and inland seas	0.008
rivers	0.0001
soil moisture	0.005
atmosphere	0.001

Major processes in the hydrologic cycle include precipitation, evaporation, transpiration, infiltration, groundwater flow, and run-off. Precipitation occurs when water vapor in the atmosphere condenses on small particles, called condensation nuclei. Evaporation is the transfer of water from the liquid to vapor state. In cold, dry conditions, ice can sublimate directly to water vapor, bypassing the liquid phase. Plants transpire water vapor through small leaf openings called stomata.

Water that reaches Earth's surface infiltrates rock and soil. The infiltration rate depends upon soil texture, degree of compaction, and ambient moisture content (Figure 1.3). Faster infiltration rates are usually associated with loose, dry, sandy soils. Compacted, wet, clay soils absorb water at a very slow rate. Worm tunnels, cracks, and other openings in soil can substantially increase infiltration rates.

Typically, soil infiltration rates are measured with a metal infiltration ring (Figure 1.4). The ring, a segment of pipe with a 15 to 30 cm diameter, is driven about 5 cm into the ground and then filled with water. Water levels are monitored over time and converted to rate values. Periodically, water is added to the ring to maintain a fairly even depth within the pipe throughout the experiment.

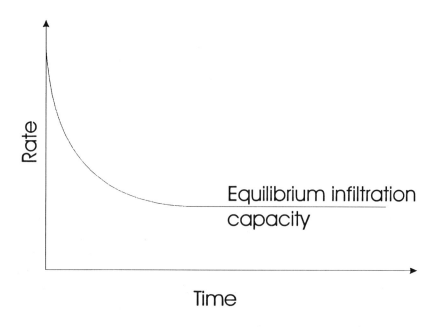

Figure 1.3. Infiltration capacity.

The most accurate infiltration rings contain an inner pipe and outer pipe. Both are filled with water, but only data from the inner pipe is used to estimate infiltration rates. The outer water is present to inhibit lateral seepage of the inner water. Such seepage would artificially increase infiltration rate estimates – during rain events, large areas of ground are flooded, and seepage is primarily vertical.

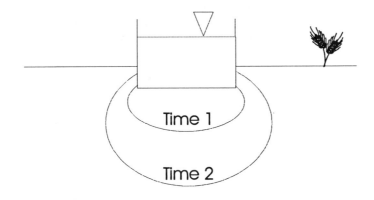

Figure 1.4. Wetting fronts at successive times for single-ring infiltrometer.

Infiltrated water may eventually reach the saturated zone, becoming groundwater. Groundwater flows through the subsurface and discharges to springs, lakes, rivers, and oceans. When groundwater flows into a river, the river is effluent or gaining (Figure 1.5). An influent or losing river is one that seeps into the ground. Effluent rivers are common in humid regions where the water table is near the land surface, such as in the southeastern United States. Rivers that are normally effluent can become influent during periods of flooding, when water levels in rivers rise much faster than groundwater. Elevated river levels cause some water to seep into the channel banks. Influent rivers are more common in arid regions, where the saturated zone is deep beneath Earth's surface.

When the infiltration capacity of the soil is satisfied, precipitation accumulates on the land surface and may eventually flow
over the land surface as runoff. This runoff can be in the form of
sheet flow, in which no channels confine the water. Much of this
sheet flow eventually reaches gullies and river channels. Rivers
drain large volumes of runoff to lakes and oceans.

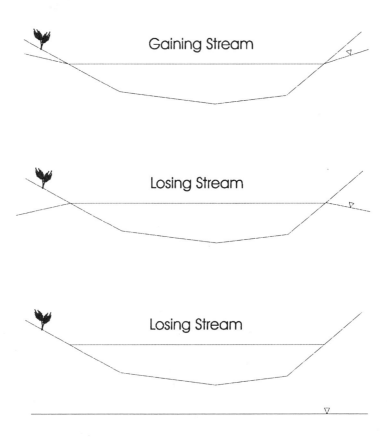

Figure 1.5. Gaining and losing streams.

Hydrologic Budget

Hydrologists often need to quantify components of the hydrologic cycle over a local study area, and then include these quantities in a hydrologic budget. Over any time period, the volume inputs (I) to a hydrologic system minus the outputs (O) must equal the change in water stored (ΔS). A system is any setting that stores water, for which water inputs and outputs can be measured over time. Lakes and wetlands are common examples.

$$I - O = \Delta S \tag{1.1}$$

For example, consider a lake that receives water from direct precipitation and stream runoff. The lake loses water by evaporation. Furthermore, it is connected to an underlying aquifer, and may gain or lose water to the groundwater domain. By measuring each component of this hydrologic system, and measuring the rise or fall of the lake level over some time period, a hydrologist could estimate the role of groundwater.

Example 1.1: A lake receives 37 inches of precipitation over a one-year period. Another 1,200 acre·ft of water enters the lake from streams. If 56 inches of water evaporated from the lake, and there were no other gains/losses to/from the lake, estimate the gain/loss from/to groundwater. The lake has a surface area of 403 acres, and its water elevation was the same at the start and end of the one-year monitoring period.

$$37 \text{ in} \times \frac{1 \text{ ft}}{12 \text{ in}} \times 403 \text{ acres} = 1{,}243 \text{ acre·ft}$$

$$56 \text{ in} \times \frac{1 \text{ ft}}{12 \text{ in}} \times 403 \text{ acres} = 1{,}881 \text{ acre·ft}$$

$$1{,}243 \text{ acre} \cdot \text{ft} + 1{,}200 \text{ acre} \cdot \text{ft} + \text{groundwater} - 1{,}881 \text{ acre} \cdot \text{ft} =$$
$$0 \text{ acre} \cdot \text{ft}$$

$$\text{groundwater} = -562 \text{ acre} \cdot \text{ft (negative sign implies lake lost this}$$
$$\text{amount to groundwater)}$$

Seepage through lakebeds or streambeds can be measured directly with a seepage meter. A seepage meter is essentially an upside-down drum or bucket pushed into the lakebed (Sanders, 1998). A tube inserted into the wall of the bucket attaches to an external plastic bag containing a known amount of water. After some time period, typically a few hours to a few days, the volume of water in the bag is re-measured. A gain or loss in volume implies the stream gained or lost water.

The seepage rate is equal to the volume change in water stored in the bag, divided by the time over which the experiment took place, divided by the cross-sectional area of the bucket where it contacts the lake bottom. To get an accurate picture of groundwater-lake interactions, seepage measurements should be taken at several places in a single lake. For example, it is possible that one side of a lake is gaining water from an aquifer while the other side is losing water.

Example 1.2: A seepage bucket is attached to a bag that receives 3 liters of water over a 9-hour period. The cross-sectional area of the seepage bucket is 0.15 m². What is the seepage rate?

$$3\,\text{L} \times \frac{1{,}000\,\text{cm}^3}{1\,\text{L}} = 3{,}000\,\text{cm}^3$$

$$\frac{3{,}000\,\text{cm}^3}{9\,\text{h}} \times \frac{1}{0.15\,\text{m}^2} \times \left(\frac{1\,\text{m}}{100\,\text{cm}}\right)^3 = \frac{0.002\,\text{m}^3}{\text{h} \cdot \text{m}^2}$$

(0.002 cubic meters per hour per square meter of lakebed)

Seepage into or out of a streambed can also be estimated by calculating the discharge of the stream at two points, one upstream and the other downstream. Stream discharge is the volume of water that flows past a channel cross-section per unit time. It is equal to the cross-sectional area of the channel times the flow velocity. Choose upstream and downstream points between which there are no tributaries contributing surface water to the stream. A higher discharge at the upstream point implies the stream is losing water to the subsurface, while a lower value at the upstream point implies gaining conditions.

Finally, water levels in nearby wells may be useful for assessing the gain or loss status of a surface-water body such as a lake or stream. Higher water levels in nearby wells (compared to the surface-water elevation) usually indicate the surface-water body is gaining groundwater. Conversely, lower groundwater levels often coincide with losing surface-water conditions.

The preceding seepage problem involved converting units. Hydrogeologists frequently convert values from one measuring unit to another – for example, from feet to meters. By using conversion factors, an initial value can be converted to an equivalent value having different units. Each conversion factor is a fraction, containing a numerator and denominator. The numerator and denominator have different units, but the entire fraction is equal to one (and therefore does not change the initial value). For example, 3.28 ft/1 m is a conversion factor equal to one. Commonly used conversion factors are listed in Table 1.2.

Example 1.3: Convert 423 gal/min to m³/day.

$$\frac{423\,\text{gal}}{\text{min}} \times \frac{1{,}440\,\text{min}}{1\,\text{day}} \times \frac{1\,\text{ft}^3}{7.48\,\text{gal}} \times \left(\frac{1\,\text{m}}{3.28\,\text{ft}}\right)^3 = \frac{2{,}308\,\text{m}^3}{\text{day}}$$

Table 1.2. Commonly used conversion factors.

Length, Area, Volume	Time	Mass
1 m = 3.28 ft	1 day = 86,400 s	1 lb = 454 g
1 mi = 5,280 ft	1 day = 1,440 min	
1 acre = 43,560 ft^2		
1 in = 2.54 cm		
1 mL = 1 cm^3		
1 darcy = 9.87 x 10^{-9} cm^2		
1 ft^3 = 7.48 gal		
1 L = 1.06 qt		

Key Terms

effluent river, evaporation, groundwater, hydrogeology, hydrologic budget, hydrologic cycle, infiltration, influent river, precipitation, runoff, saturated zone, seepage meter, sheet flow, stream discharge, sublimation, transpiration, vadose zone

Problems

1. Use the Internet or other source of information to quantify the following components of hydrologic cycle for your hometown. State your hometown in your answer.

   annual precipitation
   annual lake surface evaporation

2. What is the name of the nearest river to your hometown?

3. What are the primary sources of drinking water for your hometown?

4. What geologic conditions would favor precipitation recharging groundwater instead of evaporating or becoming runoff?

5. Suppose that you are standing on the bank of a stream. How could you tell whether the stream was influent or effluent?

6. Visit a pond or lake near where you live, and determine whether it is gaining or losing groundwater. Turn in a map showing the location of your study area, and provide your measurement methods and calculations.

7. Hydrologic data were measured at a 1,300-acre lake over a one-year time period. Estimate the volumetric contribution or loss of groundwater.

lake fell 3 inches
precipitation = 45 inches
evaporation = 59 inches
stream flow into lake = 970 acre·ft

8. The following data were recorded during an infiltration test. Plot the data and estimate the equilibrium infiltration capacity.

Infiltration Rate (cm/min)	Time (min)
1.70	1
1.14	2
0.65	3
0.43	4
0.37	5
0.35	6
0.33	7
0.32	8
0.32	9
0.31	10

9. Consider two points along a river channel, one upstream and the other downstream. The points are approximately 200 ft apart, and there are no tributaries flowing into the stream between the measurement points. At the upstream point, the channel cross-section is 4.9 m², and the water is flowing at a rate of 0.24 m/s. The channel area and flow rate are 3.1 m² and 0.42 m/s, respectively, at the downstream point. Calculate the river's discharge at each measurement point, and determine whether it is gaining or losing water between the points.

CHAPTER 2

Aquifers

Groundwater occurs within the openings of consolidated rock or unconsolidated sediment. Openings in rocks can be classified as primary or secondary (Figure 2.1). Primary openings are voids present when a rock forms. Secondary openings form afterward. Fractures and dissolution structures are common secondary openings. The interconnected passageways available for flow are usually narrow and irregular. Consequently, most groundwater flows exceedingly slow, often less than 10 m/yr. Occasionally, groundwater flows rapidly through underground caves, large fractures, or gravel deposits (Lohman, 1979).

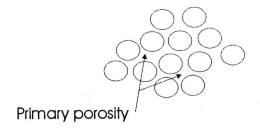

Primary porosity

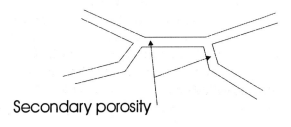

Secondary porosity

Figure 2.1. Primary and secondary porosity.

An aquifer is a body of saturated rock or sediment that is capable of transmitting useful quantities of water to wells or springs. Examples of aquifers are unconsolidated sand and gravel, sandstone, limestone, and fractured rocks. Sand and gravel can be found in alluvial valleys (Figure 2.2), tectonic valleys (Figure 2.3), and glacial outwash (meltwater) deposits. Sandstone and limestone are common sedimentary rocks, forming in various terrestrial, transitional, and marine environments. Cooling magma and tectonic forces often produce fractures in rocks. Many fractures are nearly vertical. Therefore, angled boreholes are more likely to intersect them and create a successful well (Figure 2.4) (Fetter, 1994).

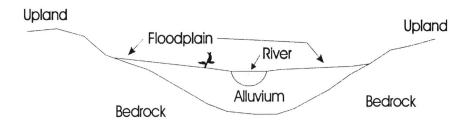

Figure 2.2. Alluvial valley.

Fractures are good conduits for migrating fluids, unless they've formed in faulted, clayey material. Displacement along fractures in clayey material forms a tight, fine-grained material known as gouge that restricts groundwater flow (Figure 2.5).

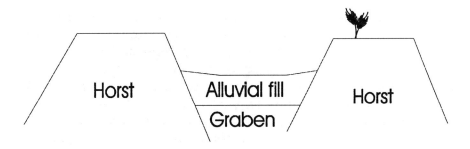

Figure 2.3. Tectonic valley.

Figure 2.4. Angled and vertical borings (bold) in fractured terrain.

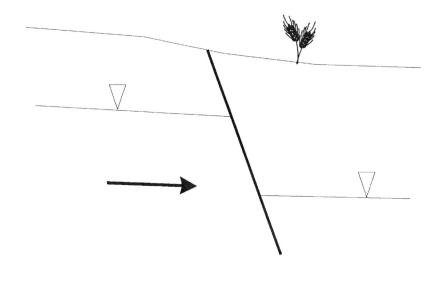

Figure 2.5. Groundwater impeded by fault gouge.

Water derived from precipitation, rivers, and lakes replenishes aquifers. For precipitation recharge to occur, there must be enough rain so that some percolates beneath the root zone. Most groundwater recharge occurs in wet seasons, when the soil is moist, and excess water seeps downward. In dry soil, water gets tied up in pores, clinging to aquifer solids. Cracks and animal burrows can facilitate groundwater recharge, because they short-circuit tortuous pathways related to primary porosity.

Water discharges from aquifers by flowing into rivers, lakes, or oceans, through some forms of vegetation, evaporation, and pumping wells. Groundwater can evaporate to the atmosphere when the water table is within a few meters of the land surface.

Confining layers, which form the boundaries of aquifers, are geologic units having a low permeability. They restrict the movement of groundwater into or out of an adjacent aquifer. An unconfined aquifer is a body of saturated sediment or rock that is not directly overlain by a confining unit (Figure 2.6). Many aquifers consist of a geologic formation that slopes into the subsurface. Such aquifers are typically unconfined in the formation's outcrop zone, but confined down slope (Figure 2.7).

The water table, a surface at which (gauge) fluid pressure equals zero, defines the top of an unconfined aquifer (Freeze and Cherry, 1979). Fluid pressure is positive below the water table and negative in the vadose zone.

Underground fluid pressure can be measured with a tensiometer (Figure 2.8) in the vadose zone or a piezometer (Figure 2.9) in the saturated zone. Tensiometers consist of a water-filled tube attached to a porous cup. Water is drawn out of the cup by suction exerted by the soil. A gauge attached to the tube measures this suction, or negative fluid pressure.

Piezometers are used extensively by hydrogeologists. This device is a solid vertical pipe, typically less than 10 cm in diameter that is open at the bottom and covered at the top with a removable cap. The bottom of the pipe is capped with a conical fitting, but just above this cap are slots or holes allowing water to enter. Lengths of piezometers range from less than one meter to several hundred meters, depending on the depth to water in a particular setting.

The pressure head (h_p) at the bottom of the pipe is equal to the length of the water column in the pipe. The elevation of the pipe's bottom, commonly reported in units of length above mean sea level, is the elevation head (h_e). Summing the elevation head and pressure head yields the hydraulic head (H) at the bottom of the pipe (Figure 2.10).

$$H = h_e + h_p \qquad\qquad\qquad (2.1)$$

In other words, the hydraulic head at a particular point in an aquifer is the elevation to which water will rise in a pipe tapping that point.

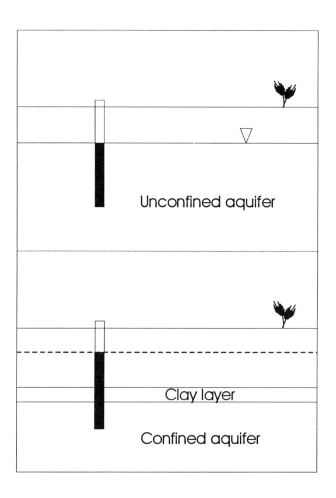

Figure 2.6. Unconfined and confined aquifers; dashed line is potentiometric surface of confined aquifer.

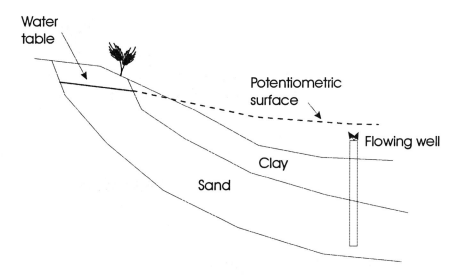

Figure 2.7. Aquifer unconfined in outcrop zone and confined downslope.

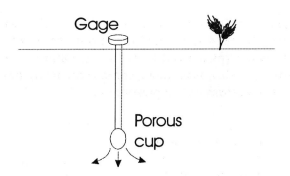

Figure 2.8. Tensiometer.

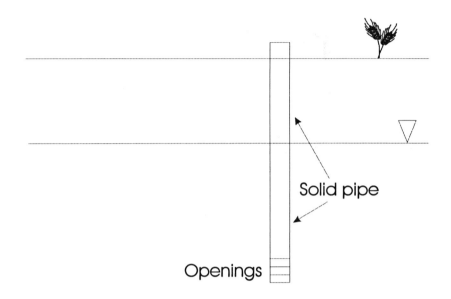

Figure 2.9. Piezometer.

Example 2.1: The top of a piezometer is 428 m above mean sea level. If the length of the piezometer is 43 m, and the depth to water (measured from the piezometer's top) is 26 m, calculate the elevation head, pressure head, and hydraulic head at the bottom of the piezometer.

$h_e = 428\,\text{m} - 43\,\text{m} = 385\,\text{m}$

$h_p = 43\,\text{m} - 26\,\text{m} = 17\,\text{m}$

$H = h_e + h_p = 402\,\text{m} \quad (\text{or } H = 428\,\text{m} - 26\,\text{m} = 402\,\text{m})$

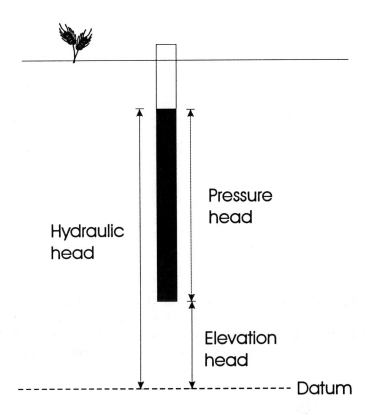

Figure 2.10. Hydraulic head, pressure head, and elevation head; darkened area represents water.

Immediately above the water table lies the capillary fringe, a saturated zone under negative fluid pressure (Figure 2.11). Within this zone, groundwater is drawn upward by adhesion of water molecules to solid particles and cohesion between water molecules. The height of capillary rise varies with sediment size. In silt, the capillary fringe may rise a few meters above the water table, and in gravel only a few centimeters.

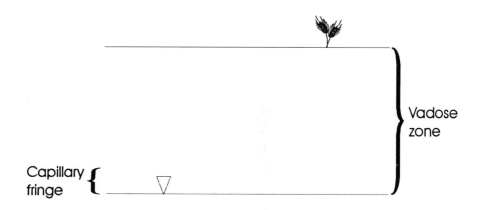

Figure 2.11. Capillary fringe.

Especially in humid regions, the water table follows the shape of the land surface, but with less relief. Occasionally, a clay layer may impede the percolation of water to the main water table. Water that accumulates on such a layer is a perched aquifer (Figure 2.12).

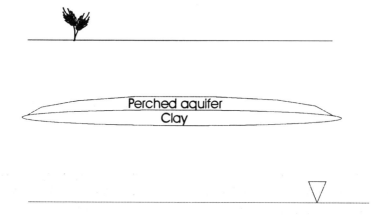

Figure 2.12. Perched aquifer.

Confining layers cap confined aquifers. The base of the cap layer defines the top of the aquifer. Groundwater that resides in a confined aquifer is under pressure. If a piezometer is placed in a confined aquifer, water from the aquifer will rise above the base of the confining layer (Figure 2.6). The potentiometric surface defines the level to which water will rise in piezometers tapping a confined aquifer. In a flowing artesian well, the potentiometric surface rises above ground level (Figure 2.7).

Physical Properties

There are several important properties that govern the capability of an aquifer to store, transmit, and yield groundwater. These include total porosity, effective porosity, intrinsic permeability, hydraulic conductivity, transmissivity, specific yield, specific retention, and the storage coefficient.

Total Porosity

Total porosity is the volume percent of a rock or soil sample that consists of void space. For most Earth material, porosity ranges from 0 to 60% (Table 2.1). It is relatively high for clay because the irregular grains do not pack well, and electrostatic charges on the surfaces of clay minerals cause particles to repel one another (Fetter, 1994). Total porosity is moderately high for most unconsolidated (loose) sediment. In contrast, it is low for sedimentary rocks because cement fills many of the pore spaces. Moreover, total porosity is extremely low in many igneous and metamorphic rocks because minerals mesh together when these rocks form.

A few generalizations can be made about total porosity values for unconsolidated sediment (Fetter, 1994). Well-sorted sediments have higher porosity values, because the particles are uniformly sized – smaller particles do not fill voids between larger particles (Figure 2.13).

Table 2.1. Representative porosity values for rocks and unconsolidated sediment (modified from Davis, 1969).

Material	Porosity (%)
Rocks	
Limestone	4.6-21.6
Chalk	29.2
Dolomite	0.4-27.8
Rock Salt	0.6
Chert	3.8
Granite	0.3
Basalt	0.8-11.4
Tuff	14-40
Pumice	87.3
Obsidian	0.52
Marble	0.3
Quartzite	0.6
Slate	3.4
Conglomerate	17.3
Sandstone	11.2-27.4
Siltstone	9.7
Shale	5.2-21.1
Sediment	
Clay	33.3-58.8
Silt	33.7-50.0
Sand	33.8-51.3

The opposite is true for poorly sorted sediments (Lohman, 1979). Well-rounded grains have lower porosity values than angular particles because they pack more tightly.

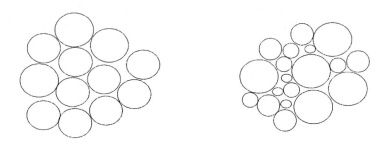

Figure 2.13. Well-sorted (left) and poorly sorted (right) sediment.

In principle, total porosity can be measured by simply pouring water from a graduated cylinder into a jar packed with a dry sample. A porosity estimate can be made based on the volume of water required to saturate the sample. However, it is often difficult to saturate fine-grained sediment and most rocks. It could take days or weeks to saturate tight clay samples having tiny pores. An alternative approach is to use the equation:

$$n = 1 - \frac{\rho_b}{\rho_s} \tag{2.2}$$

where ρ_b is the dry bulk density of a field sample (oven-dried mass divided by volume of sample) and ρ_s is the particle density (oven dried mass divided by volume of solids). A water displacement test can measure the volume of solids in a sample. This is accomplished by dropping the dry particles into a graduated cylinder containing water. The volume of solids approximately equals the change in water level within the cylinder. Note that the initial volume of water in the cylinder must be large enough so that the entire sample is under water.

Equation (2.2) can be rewritten by substituting the definitions for dry bulk density and particle density, to yield:

$$n = 1 - \frac{\text{mass/total volume}}{\text{mass/particle volume}} \qquad (2.3)$$

This reduces further to:

$$n = 1 - \frac{\text{particle volume}}{\text{total volume}} \qquad (2.4)$$

In some situations, it may be easier to calculate the porosity of a sample by simply applying Equation 2.4. In that case there is no need to weigh the sample. However, it should still be dried out before doing the water displacement test. Notice that the right-hand side of Equation 2.4 is the fraction of the total sample volume that consists of openings, or the porosity of the sample.

Example 2.2: Calculate the porosity of a sand sample given the following information.

total volume of sample when extracted from field = 188 cm^3

initial water level in graduated cylinder = 432 cm^3

displaced water level (after dried sand was submerged in cylinder) = 498 cm^3

$$n = 1 - \frac{\text{particle volume}}{\text{total volume}}$$

$$= 1 - \frac{(498\,\text{cm}^3 - 432\,\text{cm}^3)}{188\,\text{cm}^3} = 0.35$$

Effective Porosity

The volume percentage of a rock or soil sample that consists of interconnected pores through which water can flow is the effective porosity (n_e). Effective porosity cannot exceed total porosity. For unconsolidated sediments, effective porosity is nearly equal to total porosity, but it is significantly lower than total porosity for lithified rock (Fetter, 1994). This relation stems from the fact that pores in many lithified rocks are not well interconnected. For example, vesicular basalt has a total porosity above 60%, but a very low effective porosity (Figure 2.14).

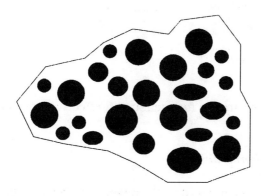

Figure 2.14. Isolated pores (black areas) in vesicular basalt.

Effective porosity can be estimated by measuring the volume of fluid required to saturate a dry sample of known volume. Typically, this is done under a vacuum to avoid air entrapment and ensure saturation. A simple laboratory apparatus can be used to estimate the effective porosity of unconsolidated sand (Figure 2.15) (Hudak, 1994a).

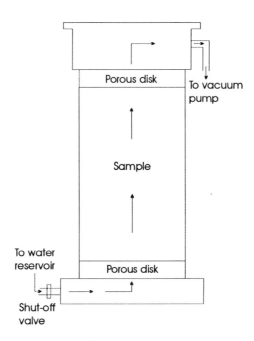

Figure 2.15. Apparatus for measuring effective porosity.

Intrinsic Permeability

Intrinsic permeability (k) measures the ability of a porous medium to transmit fluids. It is a property of the medium, independent of the fluid. The darcy, a commonly used unit of permeability, is equal to 9.87 x 10⁻⁹ cm². For unconsolidated sediment, k increases as the median grain size and degree of sorting increase.

Weathering and fracturing can increase the permeability of rocks by two to four orders of magnitude (Figure 2.16) (Davis, 1969). Both weathering and fracturing are most evident within 20 m of the land surface. In tropical regions, weathering may extend to depths of more than 100 m, but most temperate regions do not have significant weathering below about 50 m (Davis, 1969).

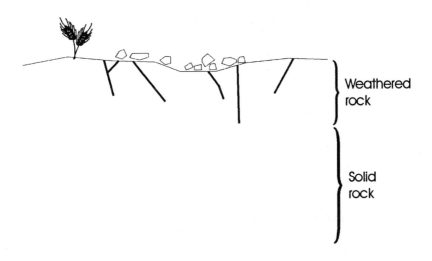

Figure 2.16. Weathered rock layer.

Underground solution enhances both the porosity and permeability of rocks (Figure 2.17). All minerals are soluble to some extent, but carbonates (such as limestone) and evaporites (such as gypsum and halite) are particularly susceptible to dissolving in water. In soluble rocks, water circulates through minor cracks, gradually widening them over time.

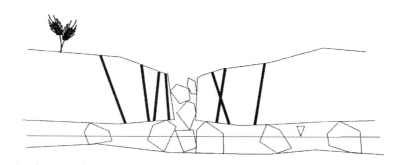

Figure 2.17. Underground solution channel.

Hydraulic Conductivity

Hydraulic conductivity (K) takes into account the permeability of an aquifer, as well as the fluid being transmitted through the aquifer. It is defined as

$$K = \frac{k\rho g}{\mu} \tag{2.5}$$

where ρ is fluid density, g is the acceleration of gravity (9.8 m/s²) and μ is fluid viscosity (g/s/cm). Shallow groundwater usually has a density and viscosity of about 1.0 g/cm³ and 0.01 g/s/cm, respectively.

Example 2.3: The intrinsic permeability of an aquifer is 2.3 x 10^{-9} cm². Calculate the aquifer's hydraulic conductivity, assuming typical values for fluid density and viscosity.

$$K = \frac{k\rho g}{\mu} = 2.3 \times 10^{-9} \text{ cm}^2 \times \frac{1.0 \text{ g}}{\text{cm}^3} \times \frac{980 \text{ cm}}{\text{s}^2} \times \frac{1 \text{ s} \cdot \text{cm}}{0.01 \text{ g}} = 2.3 \times 10^{-4} \text{ cm/s}$$

Hydraulic conductivity ranges from approximately 10^{-9} m/d for unfractured shale and igneous rocks to 1,000 m/d for gravel or cavernous limestone (Figure 2.18). Although clay has a high porosity, the pore spaces available for flow are small. Consequently, the hydraulic conductivity of clay is rather low. Hydraulic conductivity can be measured in the laboratory or field, as described in Chapters 4 and 5.

In many aquifers, K varies with direction and location. A homogeneous aquifer is one in which the magnitude of K is the same at all locations. In a heterogeneous aquifer, the magnitude of K varies from one location to another (Figure 2.19). A different term, isotropy, pertains to directional variability in the magnitude of K. Isotropic conditions mean that the magnitude of K is equal in all directions. In an anisotropic aquifer, K varies with direction (Figure 2.20). For many layered deposits, the vertical hydraulic conductivity is significantly less than the horizontal hydraulic conductivity. Vertical to horizontal hydraulic conductivity ratios of 1/10 to 1/1,000 are fairly common in sedimentary deposits. This pattern results from grains being vertically compacted and elongated in the horizontal direction. Fluids move more easily in the direction parallel to elongated grains. In contrast, horizontal and vertical hydraulic conductivity are nearly equal for spherical grains. Spherical grains might be encountered in ancient beach or dune deposits.

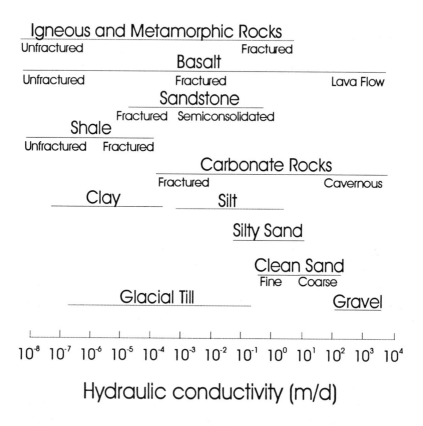

Figure 2.18. Representative hydraulic conductivity values (Heath, 1983).

Transmissivity

Transmissivity (*T*) is closely related to hydraulic conductivity. Often used for confined aquifers, transmissivity quantifies the amount of water that can be transmitted horizontally through the saturated thickness of a unit (Figure 2.21). It is equal to the product of hydraulic conductivity and saturated thickness.

$$T = Kb \qquad\qquad\qquad\qquad\qquad (2.6)$$

Example 2.4: Calculate the transmissivity of a confined aquifer having a hydraulic conductivity of 1.7 m/d and a saturated thickness of 56 m.

$$T = Kb = \frac{1.7\,\text{m}}{\text{d}} \times 56\,\text{m} = 95.2\,\text{m}^2/\text{d}$$

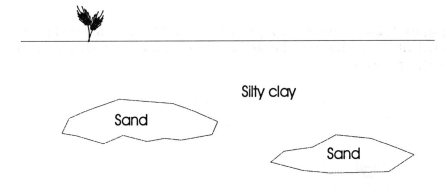

Figure 2.19. Heterogeneous aquifer.

Specific Yield

Specific yield is the volume of water drained by gravity divided by the total volume of a saturated aquifer sample. The specific yield of most unconfined aquifers ranges from about 10 to

30% (Lohman, 1979). Higher values are associated with coarser sediment (Table 2.2).

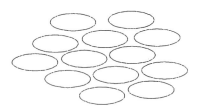

Figure 2.20. Anisotropic aquifer (in this example, hydraulic conductivity is highest in the horizontal direction and lowest in the vertical direction).

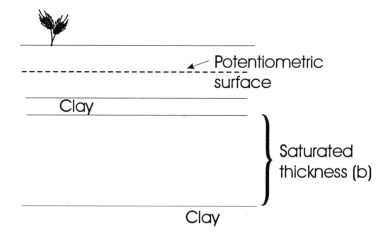

Figure 2.21. Saturated thickness.

$$S_y = \frac{V_g}{V} \tag{2.7}$$

Example 2.5: If 48 cm³ of water drains from a saturated sample having a total volume of 150 cm³, what is the sample's specific yield?

$$S_y = \frac{V_g}{V} = \frac{48 \text{ cm}^3}{150 \text{ cm}^3} = 0.32$$

Table 2.2. Representative specific yield values for unconsolidated sediment (modified from Johnson, 1967).

Sediment	Specific Yield (%)
Clay	2
Sandy clay	7
Silt	18
Fine sand	21
Medium sand	26
Coarse sand	27
Fine gravel	25
Medium gravel	23
Coarse gravel	22

Specific Retention

A saturated sample will not release all of its water by gravity drainage. Pendular water is left behind, clinging to the aquifer solids against the force of gravity (Figure 2.22). Adhesive (between water molecules and aquifer particles) and cohesive (between water molecules) forces hold this water in place. In general, fine-grained sediments, due to a larger surface area per unit volume and smaller pores, will hold more water. Specific retention is the volume of water retained against gravity divided by the volume of a sample. Specific yield plus specific retention equals the effective porosity of a sample.

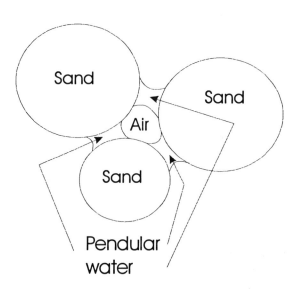

Figure 2.22. Pendular water.

$$S_r = \frac{V_r}{V} \tag{2.8}$$

Example 2.6: The specific retention of an aquifer is 0.17. If the effective porosity is 0.28, how much water will gravity-drain from 300 m³ of aquifer?

$$S_y = n_e - S_r = 0.28 - 0.17 = 0.11$$

$$V_g = S_y V = 0.11 \times 300 \text{ m}^3 = 33 \text{ m}^3$$

Storage Coefficient

During dry periods or pumping, aquifers release water from storage. Releases from storage decrease the hydraulic head in aquifers. The storage coefficient (S) of an aquifer is the volume of water released (ΔV) per unit horizontal area (A) of aquifer per unit decline in hydraulic head (ΔH).

$$S = \frac{\Delta V}{A \Delta H} \tag{2.9}$$

Storage coefficients of unconfined aquifers are virtually equal to specific yield because most of the water is released by gravity drainage (Lohman, 1979). Values for unconfined aquifers generally range from 0.1 to 0.3. In contrast, confined aquifers release water by compression of aquifer solids and water expansion. These processes occur in response to a reduction of fluid pressure within the aquifer. Gravity drainage does not apply to confined aquifers. Because confined aquifers release a small amount of water per unit decline in the hydraulic head, they have comparatively small storage coefficients, typically ranging from about 10^{-5} to 10^{-3}.

Example 2.7: The storage coefficient of a confined aquifer is 0.0005. If the hydraulic head drops 14 m over an area of 4.6 km², how much water was lost from storage?

$$\Delta V = SA\Delta H = 0.0005 \text{ x } 4.6 \text{ km}^2 \text{x} \left(\frac{1,000 \text{ m}}{1 \text{ km}} \right)^2 \text{x } 14 \text{ m} = 32,200 \text{ m}^3$$

Springs

Springs are places where groundwater seeps out at the land surface. Historically, springs have been an important source of water for humans. Our predecessors did not have the luxury of modern drilling rigs and, in the absence of surface water, they relied upon springs to supply water. Indeed, at many archaeological sites there is evidence of a nearby spring. For example, springs often deposit travertine, a freshwater limestone. Travertine deposits persist through the geologic record, even after a spring has dried up.

Fetter (1994) recognized several types of springs: depression, contact, fault, sinkhole, and fracture (joint) (Figures 2.23 and 2.24). A depression spring forms when the water table intersects a steeply sloping hillside. These springs are more common in rugged terrain than flat regions. Contact springs are created when groundwater accumulates above a layer of rock having a low hydraulic conductivity. Groundwater moves preferentially through the more permeable, overlying layer. Eventually, the groundwater discharges to a spring where the layer intersects a hillside. For example, a contact spring could form where a sandstone layer overlies a shale layer, and a road cut truncates the layers.

Faults often juxtapose permeable and impermeable rock layers. When this happens, groundwater may migrate through the permeable layer, and then travel along the fault to the land surface.

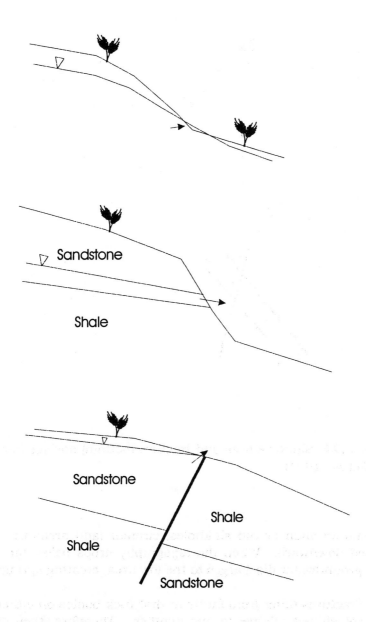

Figure 2.23. Depression (top), contact (middle), and fault (bottom) springs (modified from Fetter, 1994).

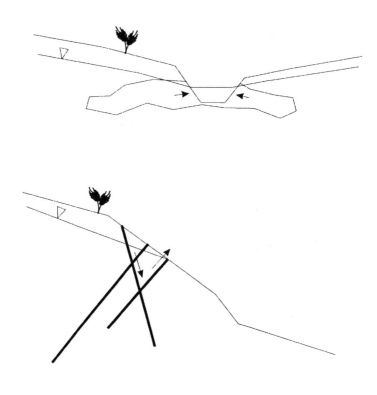

Figure 2.24. Sinkhole (top) and fracture (bottom) springs (modified from Fetter, 1994).

Steep faults often bound sinkholes, circular land areas that have dropped downward. When the topography drops below the water table, groundwater discharges to the low area, creating springs.

Fractures differ from faults in that rock bodies on either side have not shifted relative to one another. Therefore, they do not juxtapose layers of contrasting permeability, nor do they create an impermeable zone of fault gouge. In contrast, fractures create openings for groundwater. Where they intersect the land surface, groundwater may discharge to springs.

Key Terms

anisotropic, aquifer, bulk density, capillary fringe, confined aquifer, confining layer, effective porosity, elevation head, heterogeneous, homogeneous, hydraulic conductivity, hydraulic head, intrinsic permeability, isotropic, particle density, piezometer, potentiometric surface, pressure head, primary openings, secondary openings, specific retention, specific yield, springs, storage coefficient, tensiometer, total porosity, transmissivity, unconfined aquifer, viscosity, water table

Problems

1.   The intrinsic permeability of a rock is 8.0×10^{-5} darcys. Given a density of 1.0 g/cm^3 and a viscosity of 0.01 g/s/cm, what is the hydraulic conductivity of the sample?

2.   Estimate the intrinsic permeability of a rock given the following data:

     hydraulic conductivity = 105 cm/day
     viscosity = 0.01 g/s/cm
     density = 1.04 g/cm^3

3.   Consider a piezometer X. Suppose that x is a point at the bottom of the piezometer. The following measurements were obtained:

     elevation at top of piezometer (m) = 740
     depth to bottom of piezometer (m) = 150
     depth to water in piezometer (m) = 60

     a) What is the hydraulic head at point x?
     b) What is the pressure head at point x?
     c) What is the elevation head at point x?

4. What is the name of the nearest aquifer to your hometown?

5. Pumice has a high total porosity, but a low hydraulic conductivity. Explain why.

6. Estimate the total porosity of a sample given the following measurements:

    particle density = 2.67 g/cm^3
    dry bulk density = 1.50 g/cm^3

7. Why is effective porosity significantly less than total porosity for lithified rock?

8. Describe a setting in which hydraulic conductivity could be anisotropic, but also homogeneous.

9. Calculate the transmissivity of an aquifer with a hydraulic conductivity of 0.3 m/day and a saturated thickness of 12.5 m.

10. Why are storage coefficients higher for unconfined aquifers than confined aquifers?

11. A confined aquifer has a storage coefficient of 0.0003. The hydraulic head declines 10 m over an area of 100,000 m^2. How much water was released from storage?

12. Find an example of one type of spring in the region where you live. Describe the spring and its location.

13. Collect a sample of dirt from your yard and estimate its porosity. State the method you used, and provide the dirt in a sealed bag.

14. Is the hydraulic head equal to, less than, or greater than the elevation head at a point on the water table? Explain.

CHAPTER 3

Monitoring Wells

Monitoring wells yield groundwater samples and water level elevations in aquifers. They can also be used to estimate the hydraulic characteristics of aquifers. The wells are usually constructed by advancing a boring with a drilling rig, installing a well casing and screen, and backfilling the annulus between the casing and wall of the borehole (Figure 3.1).

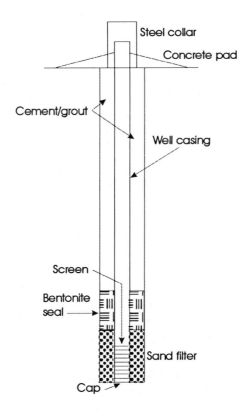

Figure 3.1. Groundwater monitoring well.

While a hole is being drilled, soil and rock samples should be collected at frequent depth intervals. Careful observations of drill cuttings can guide the collection of undisturbed samples. Soil and rock samples define the stratigraphy of a study area. In environmental investigations, these samples can be chemically analyzed to determine the extent of subsurface contamination. All cuttings and samples obtained during drilling should be carefully described in a field logbook (Figure 3.2).

Soil and rock samples are described according to several physical characteristics, including color, texture, degree of compaction, density, and moisture content. The predominant grain size names an unconsolidated sample. Other observable grain sizes may be used as adjectives. For example, a clay containing a minor fraction of sand may be described as a sandy clay. Experienced geologists can determine the texture of a sample by a field examination. Novice geologists and technicians can learn to classify textures by checking their field descriptions with jar tests.

A jar test involves submerging a soil sample in a water-filled jar. The jar is shaken vigorously, and then set aside, allowing particles to settle out of suspension. Larger particles will settle first, followed by progressively smaller particles. Clay may stay suspended for a few days. One can estimate the fraction of sand, silt, and clay by observing layers that accumulate at the jar's bottom. Sand particles will be visible to the naked eye. Although individual silt and clay particles are not visible, they typically accumulate in distinct layers that take on different colors. Many environmental firms and regulatory agencies use the Unified Soil Classification System (USCS) to describe soil texture (Figure 3.3).

The USCS classifies soil on the basis of grain size and sorting. Coarse-grained soils contain a large fraction of sand and gravel, whereas fine-grained soils contain mostly silt and clay. A well-graded soil implies a mixture of grain sizes, which could also be described as poorly sorted.

Date:	Client: Site:	Elevation at top of casing:	ID #:
Comments:			
Depth	**Sample Description**	**Completion Data**	
		Total depth: Casing size and type: Depth of screened interval: Annulus (type and depth intervals): Other:	

Figure 3.2. Well completion form.

Symbol	Description
GW	Well graded gravels, gravel-sand mixtures, little or no fines
GP	Poorly graded gravels, gravel-sand mixtures, little or no fines
GM	Silty gravels, gravel-sand-silt mixtures
GC	Clayey gravels, gravel-sand-clay mixtures
SW	Well graded sands, gravelly sands, little or no fines
SP	Poorly graded sands, gravelly sands, little or no fines
SM	Silty sands, sand-silt mixtures
SC	Clayey sands, sand-clay mixtures
ML	Inorganic silts, rock or clayey silt with low plasticity
CL	Inorganic clays of low to medium plasticity, gravelly clays, sandy clays, silty clays, lean clays
OL	Organic silts and clayey silts of low plasticity
MH	Inorganic plastic silts, micaceous or diatomaceous silts
CH	Inorganic clays of high plasticity, fat clays
OH	Organic clays of medium to high plasticity, organic silty clays
PT	Peat, humus, swamp soils with high organic content, fibrous

Note: Dual symbols are used to indicate borderline soil classifications.

Figure 3.3. Unified Soil Classification System.

Drilling Methods

The method chosen for drilling a monitoring well depends on several factors, including subsurface conditions, equipment availability, versatility of the drilling method, drilling cost, site accessibility, installation time, ability to preserve natural conditions, and ability to obtain reliable samples (EPA, 1994a). The most common drilling methods include hollow-stem auger, solid-stem auger, cable tool, air rotary, water rotary, mud rotary, dual-wall reverse circulation, sleeve- or hammer-driving, and jet percussion. In environmental investigations involving contaminated soil and rock, the drilling equipment should be steam-cleaned between holes. Cleaning pertains to all drilling methods, so that clean locations are not contaminated by dirty equipment.

Hollow-Stem Auger

The hollow-stem auger method is frequently used to install monitoring wells in unconsolidated or poorly consolidated materials, but is inappropriate for solid rock (EPA, 1994a). The auger consists of hollow, coarsely threaded pipes (Figure 3.4). A cutting head is attached to the first auger flight, and additional flights are added as the augers are rotated downward. A plug attached to a rod inside the hollow stem prevents soil from entering the interior of the stem. Cuttings move upward along the sides of the borehole to the land surface. These "disturbed samples" provide information on subsurface conditions, but cannot be assigned to a precise elevation.

Periodically, the plug is removed, and a sampling device is attached to pipe lowered down the auger to extract an undisturbed sample. Split-spoon or Shelby tube samplers are commonly used for this purpose. These tubes are usually about 46 cm long and 5 cm in diameter. A split-spoon sampler is a hollow tube comprised of two halves, whereas the Shelby sampler is a thin-walled, one-piece hollow tube. A hammering device drives the samplers into the soil ahead of the bit.

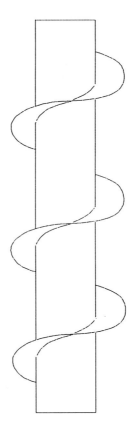

Figure 3.4. Section of hollow-stem auger.

Shelby tubes have some disadvantages over split-spoon samplers (Sanders, 1998). When a Shelby tube is used, the sample cannot be studied in the field, because the sampler is opaque. Instead, the sample and tube are packaged in the field and taken to a laboratory, where the sample is extracted for detailed study. Also, the thin walls of Shelby tubes often deform when drilling through gravel or hard clay.

The hollow stem temporarily encases the borehole, allowing the well screen and casing to be inserted once the desired depth is reached. This function is advantageous because it prevents a borehole collapse that might occur if the augers had to be withdrawn before installing the well casing. Potential disadvantages of the hollow-stem method include cross-contamination of subsurface material (drill cuttings are conveyed along the length of borehole), sand and gravel heaving into the auger, and clay smearing along the borehole wall.

Recent technological innovations have led to the development of portable hollow-stem auger rigs (H. Westmoreland, personal communication). These miniature rigs, powered by small gasoline or hydraulic motors, are capable of drilling 40 m of unconsolidated sediment or weathered rock. They employ smaller-diameter augers than the larger rigs, but an equally heavy drop hammer for driving samplers. The miniature rigs weigh less than 400 kg and can be loaded into pick-up trucks or trailers. Portable rigs can be wheeled around small, congested sites, such as gasoline stations. They are also much less expensive than larger rigs, and can be easily adapted to drilling angled holes. However, the smaller rigs are not as fast as truck-mounted rigs, nor can they efficiently penetrate hard rock.

Solid-Stem Auger

In contrast to hollow-stem augers, solid-stem augers need to be removed from the borehole to collect samples and install casing. Unconsolidated deposits that lack cohesion are not amenable to the solid-stem method. Most unconsolidated and poorly consolidated deposits have a tendency to collapse when augers are removed from the saturated zone (Figure 3.5) (EPA, 1994a). Therefore, with the solid-stem method, undisturbed samples of unconsolidated materials cannot be collected below the water table.

Figure 3.5. Slough in unsupported hole.

Bucket Auger

The bucket auger is a manual form of machine-driven augering. This drilling apparatus consists of segments of solid pipe attached to a T-shaped handle. The pipe is made of steel or aluminum, and coupled with threads or clips. Each segment of pipe is about 1.5 m long and 2.5 cm in diameter. A hollow pipe, usually about 20 cm long and 10 cm in diameter, with two cutting blades, is attached to the first segment of drill pipe. One or two people turn the handle and push downward, rotating the pipe into the ground. After about every 20 cm of penetration, the pipe must be lifted out of the hole and the bucket emptied. This method of drilling requires cohesive sediment without cobbles or boulders.

Gravel lenses or dense, dry clay may effectively terminate a drilling operation. The method is quite slow compared to alternative, automated drilling techniques. However, the bucket auger and accessories are inexpensive and easily transported over difficult terrain.

Cable Tool

Various forms of cable tool drilling have been used for about 4,000 years (Driscoll, 1986). Modern cable tool equipment can drill a well within a few days, whereas the earliest equipment often required several years. Cable tool drilling machines operate by repeatedly lifting and dropping a heavy string of drilling tools into the borehole (Figure 3.6). The drill bit breaks and crushes consolidated rock into small fragments and loosens unconsolidated material. The reciprocating action of the tools mixes the crushed or loose particles with water to form a slurry at the bottom of the hole. Water is added if none is present in the formation. Periodically, a sand pump or bailer removes the slurry.

In-situ samples can be obtained by attaching a sampling tube to the drilling string. A hammering action drives the sampling barrel into the ground. This form of sampling does not work in consolidated bedrock, but is applicable to overburden material. Borehole instability can be overcome by driving casing ahead of the sampling zone. Overall, the cable tool method is relatively inexpensive, but slow compared to rotary drilling methods.

Air Rotary

Rotary drilling involves the use of circulating fluids to remove drill cuttings and maintain an open hole as drilling progresses (Figure 3.7). Drilling proceeds much faster than in the preceding methods, because the drill stem rotates at high speeds. Borehole caving is a potential problem in unconsolidated formations. However, driving casing as the borehole is advanced can resolve this problem (Figure 3.8). In the air rotary method, air is forced down the drill pipe and back up the borehole to remove the drill cuttings. Best suited for hard rock, the air rotary method is rarely used for environmental investigations because it cannot yield representative samples (EPA, 1994a). Injecting air into a borehole may alter the natural properties of the subsurface.

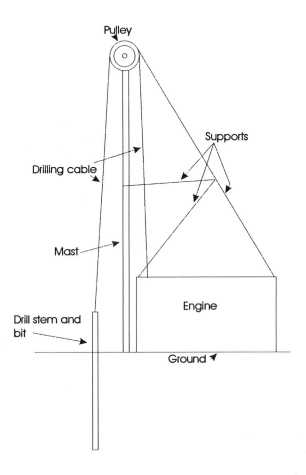

Figure 3.6. Selected components of cable-tool rig.

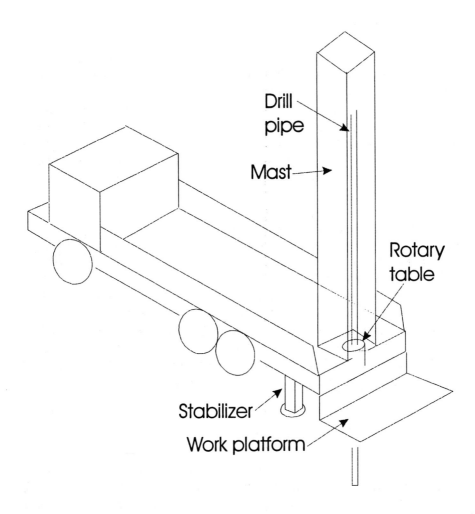

Figure 3.7. Selected components of rotary rig.

Water and Mud Rotary

Water and mud rotary drilling involves the introduction of fluids through the drill pipe to maintain an open hole, provide lubrication to the drill bit, and remove drill cuttings (EPA, 1994a). Water rotary is rapid and effective for most materials, but the fluid tends to mix with the surrounding formations and groundwater. The identification of water-bearing zones is hampered by the addition of water into the borehole. In clay-rich sediments, the water may form a slurry that plugs the formation, making it difficult to develop the well. In fractured rock, it may be difficult to maintain effective water circulation because of water losses to the subsurface.

Mud rotary is used instead of water when it is extremely difficult to maintain an open borehole. However, the additives create a high potential for affecting aquifer characteristics and groundwater quality. In environmental investigations, the mud should be limited to locally occurring clays. An additional problem with mud rotary is that cakes, which tend to form on the sides of boreholes, reduce porosity and may inhibit groundwater flow. This could prevent water from entering a well completed in the hole. Mud cakes would render a monitoring well unsuitable for sampling and severely reduce the yield of a water supply well.

Dual-Wall Reverse-Circulation

The dual-wall, reverse-circulation rotary method employs a double-wall drill pipe. Air or water is forced down the outer casing and circulated up the inner drill pipe. Cuttings are lifted upward through the pipe to the surface. Either a hammer or bit can be used to cut the formation. The primary advantage of the dual-wall method is that it allows continuous sampling of the subsurface and eliminates problems associated with lost circulation and borehole instability (EPA, 1994a). However, this method also requires relatively expensive drilling equipment and a large borehole to accommodate the dual-wall pipe.

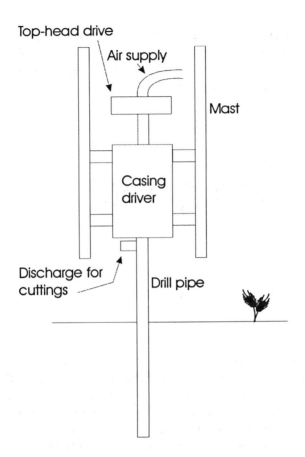

Figure 3.8. Casing driver.

Driven Wells

Driven wells consist of a steel well screen that is welded or coupled to a steel casing (Figure 3.9). A conical tip beneath the well screen penetrates soil and loose sediment. The well screen and casing are forced into the ground by hand using a weighted drive sleeve, or with a drive head mounted on a hoist (EPA, 1994a).

As the well is driven, new sections of casing are attached in 1.2-m or 1.5-m sections. Diameters of driven wells are small, most falling in the range of 2.5 to 10 cm (Todd, 1980). The method is better for shallow wells that are less than 15 m deep.

Figure 3.9. Drive-point piezometer.

It is difficult if not impossible to drive wells through dense clay, cobbles, and boulders. If driven, the well may be destroyed in these environments. In addition, silts and clays can clog the well screen. Another disadvantage of the driven-well method is that it cannot yield representative samples of materials that are penetrated. However, driven wells are less expensive than any other drilling method. Driven wells cost as little as US $5. In contrast, most rig-based methods cost at least US $1,000 per well.

Jet Percussion

The jet percussion method employs a wedge-shaped drill bit attached to the end of the drill pipe. The bit is lifted and dropped while rotating. Water is forced under pressure down the drill pipe and discharged through ports on the sides of the drill bit. It then moves up the annular space between the drill pipe and borehole wall, carrying cuttings to the surface. The jet percussion method is limited to unconsolidated or soft formations. Disadvantages include disturbing the formation and an inability to obtain representative samples (EPA, 1994a).

Well Construction

After a hole is drilled, casing and annular material must be added to complete the well. The casing consists of segments of solid pipe, whereas the screen is a slotted section of pipe. Segments of casing and screen are usually threaded, glued, or welded together. Threaded joints require a rubber O-ring to prevent leakage. Glue is less preferable than other couplers because it can leach into the groundwater. Centralizers ensure that the casing stays in the middle of the borehole (Figure 3.10). These devices usually consist of sleeves that butt against the walls of a borehole.

Figure 3.10. Well centralizers.

The well screen should be designed to allow water but not sediment to enter the well. Screen lengths vary depending on the purpose of a well, but generally should not exceed 3 m (EPA, 1994a). Long well screens provide vertically composite rather than discrete samples. However, the screened portion of a well must allow sufficient water to enter for sampling or piezometer tests.

Casing material is chosen on the basis of cost, durability, and reactivity with water. Teflon is most costly, least durable, and most inert. Stainless steel is most durable, moderate in cost, and essentially inert. PVC, which is available in most hardware stores, is least costly and most frequently used.

The annular space between the borehole wall and casing should be filled to prevent passage of formation materials into the well. Many bedrock wells do not require screens and thus do not require filter packs. However, most wells do require filter packs and a screened length of casing. Filter pack material should be chemically inert. Industrial grade silica sand or glass beads are appropriate.

The EPA (1994a) recommends that filter pack constituents be well rounded and of uniform size. The filter pack should extend at least 0.6 m above the top of the well screen. Above the filter pack, proper sealing of the annular space between the well casing and borehole wall prevents a hydraulic connection between surface water and groundwater near the well. Otherwise, surface contaminants could migrate downward, along the annulus, and pollute the well. Annular sealant should be chemically inert and essentially impermeable. A 1-m plug of bentonite is usually placed directly over the filter pack. The bentonite can be emplaced by dropping pellets down the annulus of the borehole, or by mixing slurry at the land surface and pumping it down. When properly emplaced and hydrated, the pellets form a clay plug. A mixture of cement and bentonite is often used to fill the annulus above the bentonite seal.

A narrow tremie pipe should be used to emplace bentonite and other annular material in the borehole. This strategy prevents material from bridging, or hanging up within the annulus. The tremie pipe is lowered down the annulus and gradually filled through an above-ground funnel. Gradually, the pipe is raised as filling proceeds. Once the annulus is full, the pipe is set aside for the next hole. Not using a tremie pipe could result in the absence of a sand pack around the well screen, or an inadequate bentonite seal. Either problem destroys the integrity of the well, rendering it unfit for its intended purpose.

Monitoring wells are commonly completed above-ground or flush with the ground. In either case, measures should be taken to prevent infiltration of surface runoff into the well annulus and prevent damage or vandalism of the well. Concrete should be installed at the ground surface to hold the well in place. Anchored in concrete, a locking protective casing or traffic plate should be placed around the well casing to prevent damage or unauthorized entry.

After a well is installed, it should be developed to remove any sediment blocking the well screen. This procedure improves yield and creates a well capable of producing low-turbidity samples. Turbid samples may interfere with the chemical analysis of groundwater. Well development is usually performed with a cylindrical plunging device or swab that is repeatedly lowered and raised through the water column. The plunging action forces water into and out of the well screen, thereby removing lodged silt and clay particles. Loose particles are transported into the well and removed with a bailer or pump (Bouwer, 1978). Other methods for developing a well include jetting with air or water (Figure 3.11) and over pumping. Over pumping involves pumping at a higher rate than the design rate.

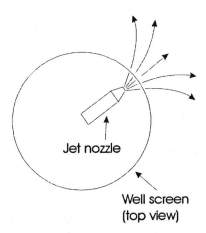

Jet nozzle

Well screen
(top view)

Figure 3.11. High-velocity jetting (modified from EPA, 1991b).

Typically, a borehole contains a single monitoring well. However, it is also possible to place multilevel sampling devices in a single borehole. These devices may consist of a series of flexible tubes tapping the sidewall of a pipe at different depths (Figure 3.12). Alternatively, multiple pipes can be placed in a single borehole, but screened at different depths (Figure 3.13). Screen lengths of 0.3 to 0.6 m are common in detailed plume geometry investigations (EPA, 1991a).

Water-Depth Measurements

Once constructed, groundwater monitoring wells should be surveyed to determine their x-y coordinates and elevation. A surveyed reference mark should be placed on the top of the well casing. This reference point should be used for all depth-to-water measurements (Figure 3.14). Its height should be determined within 3 mm or 0.01 ft in relation to mean sea level.

The depth to water in a well can be measured with various instruments, including a cloth tape and popper, a steel tape coated with chalk, an acoustic well probe, electrical sensors, pressure transducers and air lines, and float devices. Poppers are hollowed-out, metal cylinders, about 3 cm tall and 1 cm wide. They make a popping sound when dropped onto a water surface. Acoustic probes transmit sound waves from the top of a casing to the water level in a well. Electrical sensors consist of a metal probe mounted on the end of a long wire. The probe is lowered down the well and sets off a light or sound when submerged in water.

Pressure transducers and air bubblers measure the height of the water column above a probe near the bottom of a well. Dataloggers attached to pressure transducers store water level data for future use. The data can be downloaded directly to a laptop or desktop computer. Pressure transducers are especially useful when water levels are changing rapidly, such as during aquifer tests (discussed in Chapter 5). Finally, float devices rest directly on the water column, recording fluctuations over time.

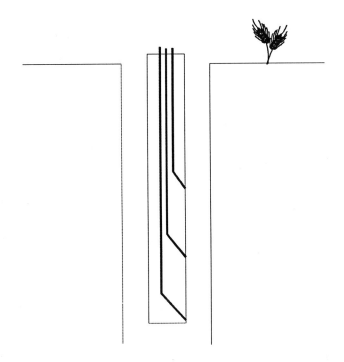

Figure 3.12. Multiple tubes tapping wall of PVC pipe in single borehole.

Regardless of the equipment used, the depth to water should be measured to the nearest 3 mm. Measuring devices should be chemically inert and not prone to chemical sorption (attraction of chemicals from water to measuring device) or desorption (removal of chemicals from measuring device to water). Instruments should be thoroughly decontaminated after each measurement to prevent cross-contamination of the groundwater. It is good practice to take the first measurement at the least contaminated well. Subsequent measurements should be made at progressively more contaminated wells.

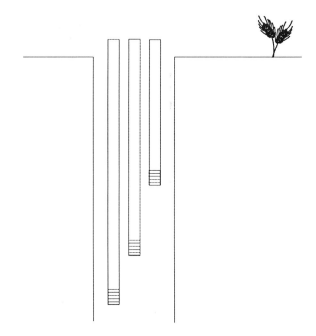

Figure 3.13. Multiple piezometers in single borehole.

At sites contaminated by leaking underground fuel tanks, there may be a layer of gasoline "floating on" the water column in a monitoring well. Interface probes can be lowered down a well to measure the depth to gasoline, and also the depth to the gasoline-water boundary. However, when using depth-to-water data at a contaminated well, Equation 3.1 should be used to correct for the presence of gasoline (Testa and Pacskowski, 1989). The gasoline "depresses" the water column from its true, undisturbed level.

$$\text{CDTW} = \text{DTW} - (\text{PT} \times G) \qquad\qquad (3.1)$$

where CDTW is the corrected depth to water, DTW is the measured depth to water, PT is the product (gasoline) thickness, and G is the specific gravity of the product. Specific gravity is defined as

Site Name:			
Date:		Logger:	
Location	Time	Depth to Water	Comments

Figure 3.14. Water-level recording form.

$$G = \frac{\rho_g}{\rho} \hspace{5cm} (3.2)$$

where ρ_g is the density of the product and ρ is the density of water. For gasoline, G is less than 1.0 and typically around 0.8.

Example 3.1: Compute the corrected depth to water in a monitoring well given the following information:

depth to gasoline = 15.84 ft
depth to gasoline-water interface = 18.45 ft
specific gravity of gasoline = 0.78

CDTW = 18.45 ft − (2.61 ft x 0.78) = 16.41 ft

Monitoring Programs

An important function of groundwater monitoring wells is supplying samples for evaluating water chemistry. Groundwater monitoring programs can be classified as ambient, source, case preparation, or research monitoring (Barcelona and others, 1984).

Ambient monitoring involves establishing conditions within a regional setting. Samples are collected routinely, over a period of many years, to determine changes in water quality over time. Generally, these samples come from a variety of public and private water supply wells. Federal and state agencies, such as the Environmental Protection Agency, U.S. Geological Survey, and Texas Water Development Board, often conduct ambient monitoring. These agencies maintain databases for thousands of wells covering several years of measurements. Individual wells may be sampled yearly, or perhaps less frequently depending largely on budget constraints.

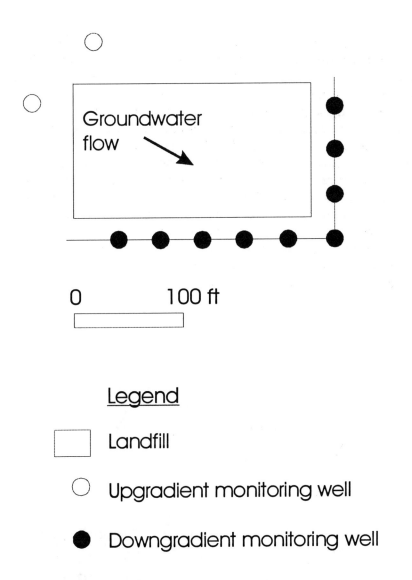

Figure 3.15. Wells immediately adjacent to landfill (modified from EPA, 1994a).

Source monitoring, also called regulatory or detection monitoring, is conducted in clean aquifers at risk of contamination from a waste storage facility. The purpose of a detection monitoring system is to identify groundwater contaminants before they traverse a downgradient regulatory boundary (Hudak, 1994b). Effective detection monitoring networks can prevent widespread degradation of groundwater supplies by enabling early aquifer remediation. A common problem is placing detection wells too close to a landfill (Figure 3.15). Contaminant plumes are narrow early on, and difficult to detect. There is some advantage to placing monitoring wells further downgradient from a landfill, for example, 20 m away (the setback depends on site specific conditions) (Figure 3.16). However, placing detection wells arbitrarily far away from a landfill is inappropriate, because by the time they pick up the contamination, a large volume of aquifer will be polluted.

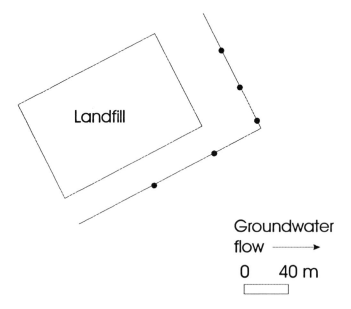

Figure 3.16. Downgradient detection monitoring network set away from immediate edge of landfill (Hudak, 1998). Reprinted with permission from Marcel Dekker.

Table 3.1. Factors affecting number of wells per location (clusters) (EPA, 1994a).

One Well Per Sampling Location[1]	More Than One Well Per Sampling Location
* No LNAPLs or DNAPLs (immiscible liquid phases)	* Presence of LNAPLS or DNAPLS

* Thin flow zone (relative to screen length)
* Horizontal flow predominates
	* Thick flow zones
	* High vertical gradient present
	* Heterogeneous, anisotropic uppermost aquifer; complicated geology
	  - multiple, interconnected aquifers
	  - variable lithology
	  - perched water zones
	  - discontinuous structures

Homogeneous, isotropic uppermost aquifer; simple geology	* Discrete fracture zones in bedrock * Solution conduits (i.e., caves) in karst terranes * Cavernous basalts

1. At the majority of sites, well clusters will be necessary to establish vertical hydraulic gradient and the vertical distribution of contaminants.

Adequate coverage in vertical as well as horizontal space should characterize a detection-monitoring network. For this reason, clusters or nests of monitoring wells are often used. Each nest has multiple sampling zones, tapping different vertical intervals (Table 3.1).

Vadose zone monitoring should be used in conjunction with groundwater detection wells (Cullen and others, 1995). Devices that can provide an early warning of contaminants in the vadose zone include moisture probes deployed in vertical and horizontal access tubes beneath the leachate collection system of a landfill, soil gas sampling points in gravel-filled trenches beneath the subgrade, and pore-liquid samplers that collect water passively or by suction (Figure 3.17). Vadose zone wells are similar to the well depicted in Figure 3.1, but they do not reach the water table. In general, vadose zone sampling devices should be installed before the footprint of a landfill is constructed.

Monitoring for case preparation includes documenting environmental contamination near a waste repository. Often the objective is to characterize the extent of contamination and identify the responsible party. This is usually an expensive endeavor, requiring that numerous wells be installed near a probable source of contamination.

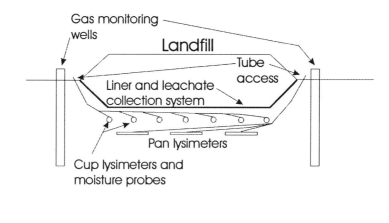

Figure 3.17. Vadose zone monitoring network.

Geophysical methods such as electrical resistivity can help guide drilling efforts (Figure 3.18). During an electrical resistivity test, a current is transmitted through the ground, and the ability of the subsurface to transmit the current is measured. This ability to conduct (or resist) an electrical current depends on such factors as sediment texture, water content, and the presence of contaminants.

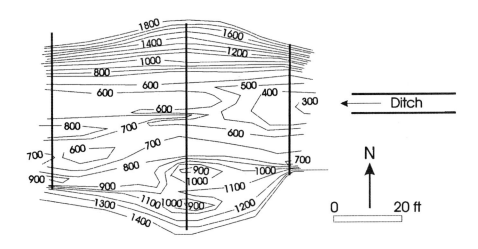

Figure 3.18. Apparent resistivity contours (ohm-ft) in area impacted by oil well brine (Hudak, 1997a). Reprinted with permission from the American Institute of Professional Geologists.

Finally, research monitoring is conducted to gain new insight on the processes that govern the movement of groundwater and the constituents it carries. This type of monitoring usually requires a high level of detail and sophistication. A good example of research monitoring is a tracer test, in which dyes are released underground and tracked with monitoring devices. The results are mapped and studied to learn how groundwater behaves in certain media.

Key Terms

air rotary, ambient monitoring, bucket auger, cable tool, case-preparation monitoring, casing, driven wells, dual-wall reverse-circulation, filter pack, hollow-stem, jet percussion, monitoring well, multilevel sampling, research monitoring, screen, Shelby tube, solid-stem, source monitoring, specific gravity, split-spoon, tremie pipe, vadose zone monitoring, water and mud rotary

Problems

1. What drilling method(s) would be suitable for installing a 30-ft deep groundwater monitoring well in an unconfined, clayey sand aquifer?

2. Describe an ideal situation for using driven wells.

3. Choose a landfill in the region where you live, and illustrate its detection monitoring system. Show the well locations, and report how often they are sampled.

4. Describe a situation that might warrant case preparation monitoring.

5. In general, how would the layout of a detection monitoring network differ from that of a case preparation monitoring network at a waste storage facility?

6. Correct a depth-to-water measurement of 8.67 ft for the presence of gasoline. Use a G value of 0.80 and each of the following product thickness values.

    a. product thickness = 1.00 ft
    b. product thickness = 2.00 ft
    c. product thickness = 3.50 ft

CHAPTER 4

Groundwater Flow

Hydraulic head, hydraulic conductivity and effective porosity control the flow of groundwater, which generally flows in the direction of the steepest hydraulic gradient.

Hydraulic Gradient

The hydraulic gradient between two points in an aquifer is the difference of the hydraulic head divided by the distance between the points (Figure 4.1).

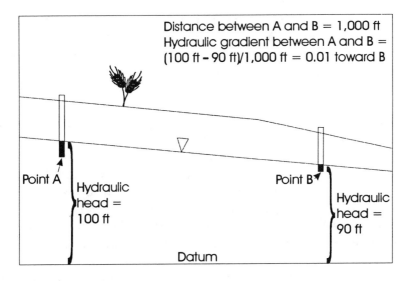

Figure 4.1. Calculation of hydraulic gradient.

At least three wells are required to establish the local direction of horizontal groundwater flow in an aquifer. With three wells, hydrogeologists can only define a planar hydraulic head surface and a single flow direction (along the plane's steepest gradient). In reality, the flow direction may vary with location in a study area, but more than three wells would be necessary to characterize the different flow directions. A rapid determination of a single flow direction can be made using the three-point method (Figure 4.2).

On a scaled map, construct a triangle using the three wells as vertices. Next, identify the side of the triangle connecting the highest and lowest water level elevations. Find the point on this line having an elevation equal to that measured in the third well. This can be accomplished by subtracting the lowest water level elevation from the intermediate value, and dividing the result by the difference between the highest and lowest water level elevations. Multiply that quotient by the length of the line. The result is the distance along the line, measured from the well with the lowest water level elevation, at which the intermediate elevation is inferred. Connect that point with the third well. The direction of groundwater flow is perpendicular to that line, in the direction of the decreasing hydraulic head.

By drawing the perpendicular to the point with the lowest water level elevation, the magnitude of the horizontal hydraulic gradient can be estimated. Measured in the direction of groundwater flow, the hydraulic gradient is the difference of the hydraulic head divided by the distance between two points. Subtract the lowest water level elevation from the intermediate value, and divide that difference by the length of the perpendicular. The result is the hydraulic gradient, expressed as a dimensionless quantity.

Groundwater flow directions should be determined from water levels measured in wells screened in the same hydrostratigraphic position (Figure 4.3). In heterogeneous geologic settings, long well screens can intercept stratigraphic intervals with different groundwater flow directions and different hydraulic head values.

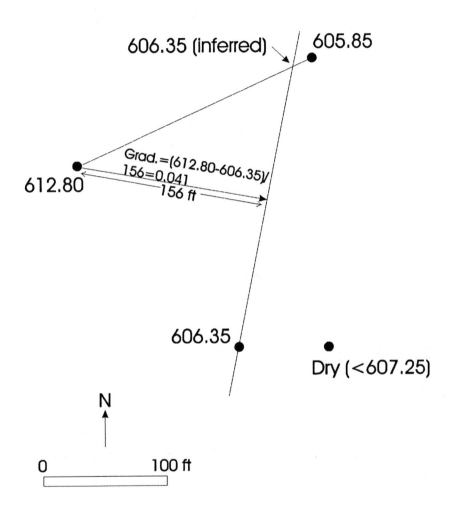

606.35 (inferred) 605.85

Grad.=(612.80-606.35)/
156=0.041
156 ft

612.80

606.35

Dry (<607.25)

N

0   100 ft

Figure 4.2. Three-point method.

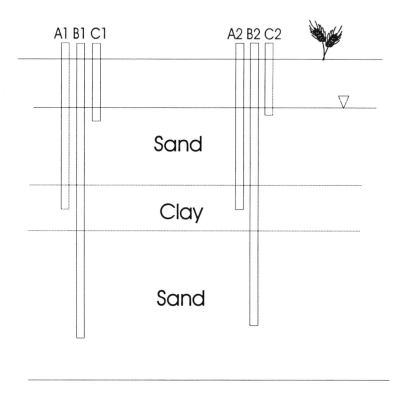

Figure 4.3. Wells A, B, and C sample different water-bearing formations.

Groundwater Velocity

The hydraulic gradient (i), in conjunction with hydraulic conductivity and effective porosity, can be used to estimate the groundwater velocity in an aquifer. Groundwater velocity (v) is defined as

$$v = \frac{Ki}{n_e} \tag{4.1}$$

Example 4.1: Calculate the groundwater velocity in an aquifer having a hydraulic gradient of 0.003, a hydraulic conductivity of 0.01 m/d, and an effective porosity of 0.31.

$$v = \frac{Ki}{n_e} = \frac{0.01\,\text{m}}{d} \times 0.003 \times \frac{1}{0.31} = 9.7 \times 10^{-5}\ \text{m/d}$$

Darcy's Law

Equation 4.1 expresses the average linear rate at which groundwater moves through an aquifer. The volumetric rate of flow can be obtained by

$$Q = KiA \tag{4.2}$$

where A is the cross-sectional area of an aquifer, measured perpendicular to groundwater flow. This area includes both aquifer solids and pores. Equation 4.2 is known as Darcy's law, in honor of the French engineer, Henry Darcy, who derived the equation while passing water through a sand-filled tube in a laboratory (Darcy, 1856).

Example 4.2: How much groundwater flows through a 435 m² section of an aquifer (oriented perpendicular to groundwater flow), under a hydraulic gradient of 0.02 and a hydraulic conductivity of 1.6 m/d?

$$Q = KiA = \frac{1.6\,\text{m}}{d} \times 0.02 \times 435\,\text{m}^2 = 13.9\ \text{m}^3/\text{d}$$

Darcy's law has many applications in hydrogeology. One is measuring the hydraulic conductivity of a sample in the laboratory. For granular material such as sand, this measurement is accomplished with a constant-head permeameter (Figure 4.4). Initially, the sample must be saturated, from the bottom to top, preferably using a vacuum to avoid air entrapment. Water passes through the device at a steady rate – the water level in the upper reservoir is fixed, and inflow to the chamber equals outflow.

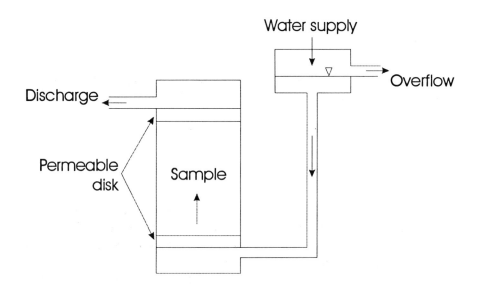

Figure 4.4. Constant-head permeameter.

Hydraulic conductivity is calculated by solving for K in Equation 4.2. Q is equal to the volumetric flow rate at the outlet, i is the hydraulic gradient between the top and bottom of the sample, and A is the cross-sectional area of the sample chamber. The hydraulic gradient is equal to the difference in water levels between the reservoir and outlet divided by the length of the sample. The experiment should be set up to avoid a large gradient that induces turbulent flow – Darcy's law is valid only for laminar flow. The hydraulic gradient should not exceed 0.5. Also check the permeameter for leaks during the test. Leaks will yield an artificially high K value – they must be repaired and the test repeated.

> **Example 4.3:** In a constant-head permeameter test, 238 cm³ of water was discharged over 190 seconds. The cross-sectional area of the chamber was 35 cm², and the sample was 19 cm long. Finally, the hydraulic head differential between the inlet and outlet was 6 cm. Estimate the hydraulic conductivity of the sample.

$$K = \frac{Q}{iA} = \frac{238\,\text{cm}^3}{190\,\text{s}} \times \frac{19\,\text{cm}}{6\,\text{cm}} \times \frac{1}{35\,\text{cm}^2} = 0.11\,\text{cm/s}$$

It is difficult to induce steady flow through fine-grained sediment. For material that transmits groundwater at a slow rate, a falling-head permeameter is used to measure K (Figure 4.5). The experiment is performed in a similar fashion to the constant-head test, except that water is allowed to fall within a tube-reservoir over a period of several hours. The falling-head equation is

$$K = \frac{A_t L}{A_c t} \ln \frac{h_i}{h_f} \qquad (4.3)$$

where t is the time length of the experiment, h_i is the initial water level above the outlet, h_f is the final water level above the outlet after time t, L is the length of the sample, A_t is the cross-sectional area of the falling-head tube, and A_c is the cross-sectional area of the sample chamber. A falling head experiment must be completed before h_f reaches zero.

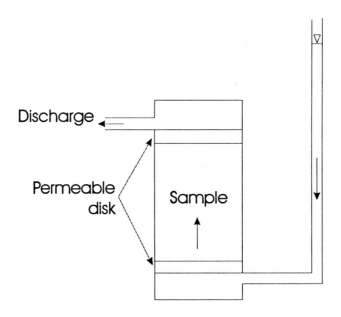

Figure 4.5. Falling-head permeameter.

Example 4.4: Compute a hydraulic conductivity value from the following data collected during a falling-head permeameter test.

$$L \ = 14 \text{ cm}$$
$$A_t = 1.8 \text{ cm}^2$$
$$A_c = 31 \text{ cm}^2$$
$$t \ = 2.35 \text{ days}$$

$$h_i = 18 \text{ cm}$$
$$h_f = 13 \text{ cm}$$

$$K = \frac{A_t L}{A_c t} \ln \frac{h_i}{h_f} = 1.8 \text{ cm}^2 \text{ x } 14 \text{ cm x } \frac{1}{31 \text{ cm}^2} \text{ x } \frac{1}{2.35 \text{ d}} \text{ x } \ln\left(\frac{18 \text{ cm}}{13 \text{ cm}}\right)$$

$$= 0.11 \text{ cm/d}$$

Flow Nets

Flow nets graphically illustrate the movement of groundwater in two-dimensional space. A flow net consists of equipotential lines and flow lines. Equipotential lines (sometimes called hydraulic head contour lines) connect points of equal hydraulic heads. Flow lines depict (simplified) paths followed by molecules of water as they move through the aquifer.

In an isotropic aquifer, groundwater flows in the direction of the steepest hydraulic gradient, and flow lines cross equipotential lines at right angles. By convention, flow nets are constructed such that the intersections between pairs of equipotential lines and flow lines form approximately square elements (Figure 4.6).

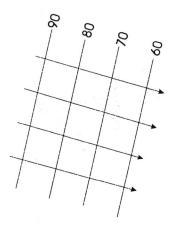

Figure 4.6. Flow net.

Flow nets can be constructed either in a horizontal plane, or a vertical plane that is parallel to groundwater flow (Figure 4.7). Horizontal flow nets are often referred to as water table or potentiometric surface maps. There can be multiple horizontal flow nets for an area, each representative of a different vertical interval within an aquifer. Generally, the uppermost interval of such a sequence would correspond to the water table (Figure 4.8).

In a water table map (unconfined aquifer), contours point upstream when they cross a gaining stream and downstream when they cross a losing stream (Figure 4.9). At a groundwater recharge area, water flows downward and away from a high area in the water table. On a flow net, diverging flow lines depict recharge areas. Groundwater discharge areas occupy low areas of the water table, often occurring at rivers or lakes. Groundwater flows toward discharge areas, which are depicted by converging flow lines on flow nets.

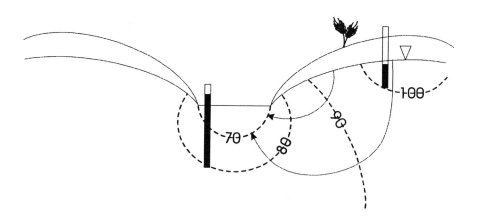

Figure 4.7. Vertical flow net.

In addition to converging flow lines (as viewed in a horizontal plane), discharge areas are often associated with upward ground-water flow. This is especially true of regional flow paths that end at a surface water body (Figure 4.7). These flow paths may penetrate deeply into an aquifer, well below the elevation of a lake or river, and then rise upward as they approach the surface water body.

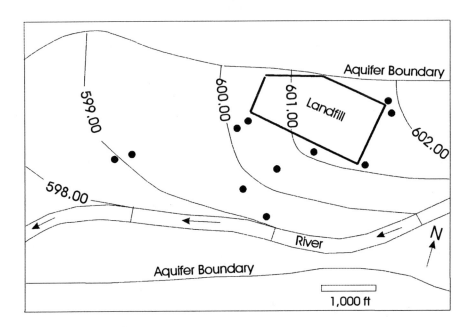

Figure 4.8. Water table contour map for site in southwest Ohio, USA; contours in feet above mean sea level (Hudak and others, 1993). Reprinted with permission from the American Water Resources Association.

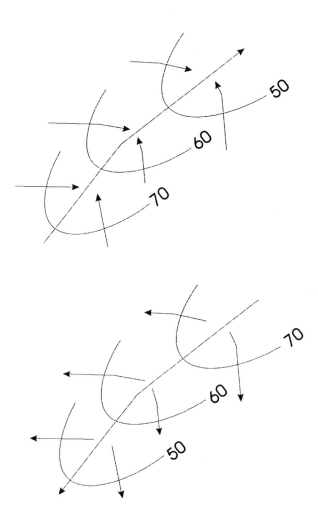

Figure 4.9. Horizontal flow net near gaining stream (also a discharge zone) (top) and losing stream (also a recharge zone) (bottom).

In a vertical section or side view, the water level in a pie-zometer will be higher than the stream level when the stream is gaining, and lower than stream level under losing conditions (Figure 4.10). Furthermore, the hydraulic head along the channel bottom is equal to the stream level. These same relationships hold for a lake. Also recall that at the water table, the pressure head is zero, and therefore the hydraulic head is equal to the elevation head.

Potentiometric surface maps for confined aquifers usually show less relation to surface topography, because they are based on piezometers tapping deeper intervals. These maps require that wells are open only to the specific aquifer of interest. For a water table map, well screens should traverse the water table, or be located just beneath the water table.

Water level measurements used to define a single water table or potentiometric surface should be made within a time interval less than 24 hours. A shorter time interval may be necessary if water levels are changing due to natural or artificial stresses, such as recharge from precipitation or pumping.

Flow Net Boundaries

Boundaries define the edges of flow nets. An arbitrary boundary is one that does not coincide with any particular feature of a flow system. The hydraulic head is variable along such a boundary. The water table is also a variable hydraulic head boundary, along which the pressure head is zero.

In contrast, a constant hydraulic head boundary implies a single hydraulic head value. For example, the level of a regulated reservoir to which an aquifer discharges may be relatively constant over time (Figure 4.11). Groundwater flows perpendicular to constant hydraulic head boundaries (Figure 4.12). A no-flow boundary cannot be traversed by groundwater. Therefore, it runs parallel to groundwater flow and perpendicular to equipotential lines (Figure 4.12).

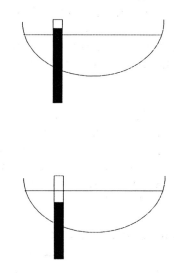

Figure 4.10. Hydraulic head levels beneath gaining (top) and losing (bottom) streams.

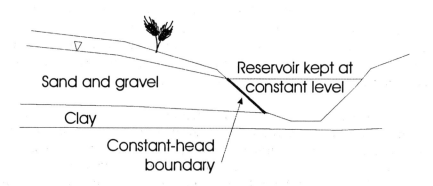

Figure 4.11. Constant-head boundary.

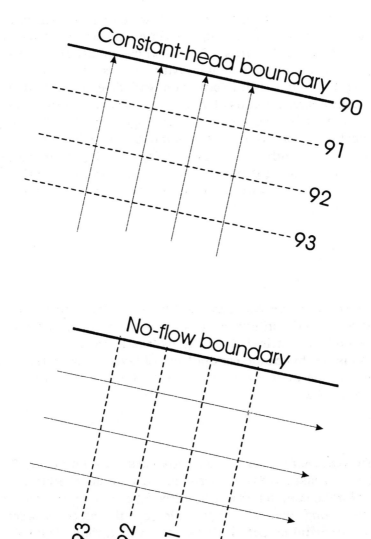

Figure 4.12. Pattern of flow lines near constant-head and no-flow boundaries.

The direction that groundwater flows changes when it passes
from one unit to another having a different hydraulic conductivity
value. In general, flow paths refract to show an affinity for material
having a higher hydraulic conductivity. Hence, groundwater that
is flowing through a sand formation will diverge around a clay
lense. Conversely, groundwater moving through a sandy clay aquifer will refract toward a gravel body (Figure 4.13). This implies
that detection monitoring wells should be screened in the more
permeable lenses within a complex geologic setting, to increase the
chance of intercepting contamination. The hydraulic conductivity
of adjacent formations dictates the degree of flow line refraction,
according to Equation 4.4 (Figure 4.14):

$$\frac{\tan \theta_1}{\tan \theta_2} = \frac{K_1}{K_2} \qquad\qquad (4.4)$$

where θ_1 is the angle between the incoming flow line (in layer 1)
and a vertical to the interface, θ_2 is the angle between the outgoing
flow line and the vertical, K_1 is the hydraulic conductivity of layer
1, and K_2 is the hydraulic conductivity of layer 2. Note that a flow
line meeting the interface head on (so that θ_1 equals zero) will not
refract in layer 2.

Example 4.5: A flow line approaches an interface at 34 degrees (measured from a perpendicular to the interface). If the
hydraulic conductivity of the first layer is 7 m/d, and that of
the second layer is 1 m/d, calculate the angle between the
flow line and perpendicular as it travels through layer 2.

$$\frac{\tan \theta_1}{\tan \theta_2} = \frac{K_1}{K_2} \Rightarrow \theta_2 = \tan^{-1}\left(\frac{K_2 \tan \theta_1}{K_1} \right) = 5.5 \text{ degrees}$$

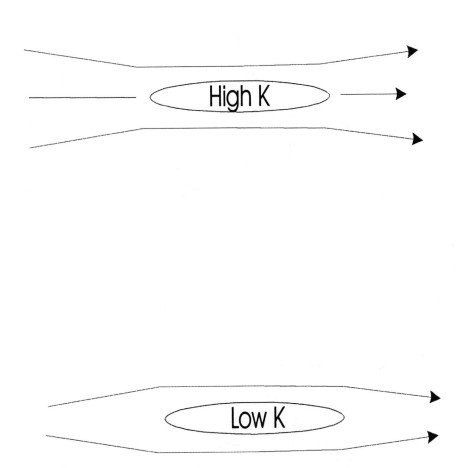

Figure 4.13. Flow-line refraction near high-*K* and low-*K* lenses.

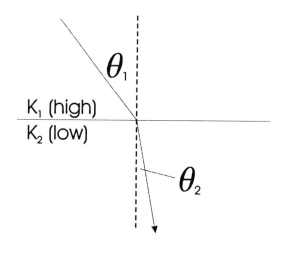

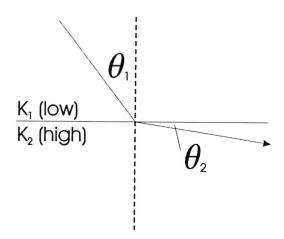

Figure 4.14. Flow-line refraction across layer interfaces.

Key Terms

arbitrary boundary, constant-head boundary, constant-head permeameter, Darcy's law, discharge area, equipotential line, falling-head permeameter, flow line, flow net, groundwater velocity, hydraulic gradient, no-flow boundary, potentiometric surface map, recharge area, refraction, variable-head boundary, water table map

Problems

1. Estimate a hydraulic conductivity value using the following data from a constant-head permeameter test:

    length of sample = 24 cm
    cross-sectional area of sample chamber = 31 cm²
    head differential between reservoir and outlet = 5.0 cm
    400 mL water collected at outlet in 17 min

2. Estimate a hydraulic conductivity value from a falling-head permeameter experiment given these parameters:

    diameter of falling-head tube = 2.0 cm
    diameter of chamber = 15 cm
    length of sample = 20 cm
    initial water level above outlet = 5.0 cm
    final water level above outlet = 1.5 cm
    duration of experiment = 18 hrs

3. Solve a three-point problem for the monitoring wells in Figure 4.15. Draw an arrow showing the inferred direction of groundwater flow. Express this direction in degrees (0° = N; 90° = E; 180° = S; 270° = W). Calculate the horizontal hydraulic gradient. Assuming a hydraulic conductivity of 1.5 ft/day and an effective porosity of 0.30, what is the groundwater velocity? Under what conditions would a three-point approach to calculating the hydraulic gradient and groundwater velocity be inappropriate?

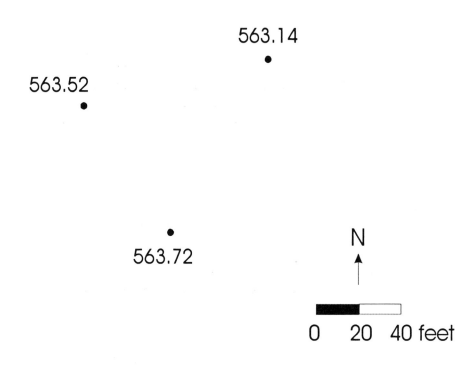

Figure 4.15. Three-point problem; water table elevations in feet above mean sea level.

4.   Figure 4.16 shows several groundwater monitoring wells and the water table elevation recorded at each well. Construct a water table contour map on the figure. Use contours of 50, 60, 70, 80 and 90 ft. Label each contour line with the appropriate numeric value. Draw flow lines on the map to construct a flow net. Label the recharge and discharge areas. Is the river gaining or losing?

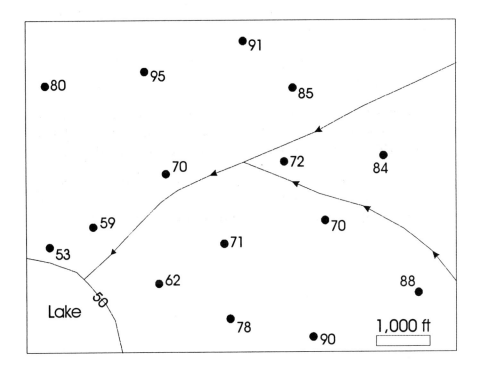

Figure 4.16. Water table elevations; lines with arrows represent rivers.

5. Choose any flow line on your flow net, and highlight it with a marker. Select two points along the line (label them "A" and "B"), and calculate the hydraulic gradient between them. Specify both the magnitude and direction (i.e., from A to B, or B to A) of the hydraulic gradient. Express the gradient as a dimensionless value. Also calculate the groundwater velocity

between points A and B. Assume a hydraulic conductivity of 5 ft/d and an effective porosity of 0.30.

6.   Label the arbitrary and constant hydraulic head boundaries in Figure 4.16.

7.   Calculate the volumetric rate of groundwater flow in an aquifer with a cross-sectional area of 1.4 km², hydraulic gradient of 0.005, and hydraulic conductivity of 10 m/d.

8.   Draw a vertical flow net for the aquifer beneath the impermeable dam in Figure 4.17.   The upstream and downstream lake levels are 456 m and 440 m above mean sea level, respectively.

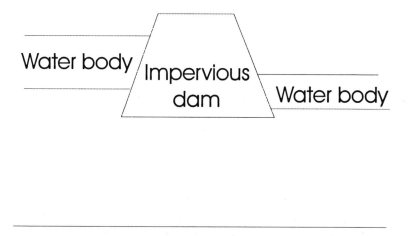

Figure 4.17. Impervious-dam problem.

9. A flow line approaches an interface between two layers at 18 degrees. If the hydraulic conductivity of the first layer is 0.8 m/day, and that of the second layer is 1.5 m/day, what is the angle of refraction in layer 2?

CHAPTER 5

Well Hydraulics and Aquifer Tests

Groundwater wells have several purposes, including agricultural, domestic, municipal, and industrial supply, removing contaminated water from aquifers, dewatering for construction, relieving pressures under dams, draining farmland, disposing of wastewater, and artificially recharging aquifers.

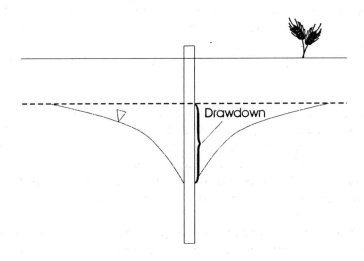

Figure 5.1. Depressed water level at pumping well.

Withdrawing groundwater from aquifers depletes and depresses water levels around the pumping well (Figure 5.1). The water table (or potentiometric surface) that forms around the well is called a cone of depression (Figure 5.2). As pumping continues, new hydraulic gradients may either reduce the aquifer's discharge or induce recharge (Sun, 1986). The cone grows and water levels decline until the rate of flow into the cone equals the pumping rate. For example, this can happen if the cone intercepts enough of the flow in the aquifer to equal the pumping rate, intercepts a stream or lake, receives recharge from precipitation, or receives leakage from other formations.

When pumpage is greater than the amount of water that can be obtained by increases in recharge or reductions in discharge, sustained withdrawal from aquifer storage and continued lowering of water levels occur. This increases pumping costs, decreases well yields, and depletes groundwater resources. Sustained water level declines may also degrade water quality, and even lower the land surface if the aquifer contains compressible clay and silt (Sun, 1986).

Steady flow prevails when the cone stabilizes and water levels are no longer dropping. In a stabilized cone of depression, hydraulic gradients are progressively smaller away from the well because the flow is moving through a larger area of the aquifer. Given a larger cross-sectional area of flow, relatively small hydraulic gradients are required to yield a particular discharge.

There are essentially two approaches to studying well hydraulics. The direct approach involves predicting water level drawdown given values for aquifer parameters such as hydraulic conductivity. Conversely, the inverse problem is to predict values for aquifer parameters given observations of drawdown. Well hydraulics equations are used for both approaches.

The equations that follow assume a constant pumping rate, horizontal flow unimpeded by aquifer boundaries, homogeneous and isotropic conditions, and insignificant well storage (Lohman, 1979; Heath, 1983). A pumping well that is screened across the entire thickness of an aquifer is necessary for horizontal flow. Otherwise, the pumping action will induce vertical hydraulic gradients and upward or downward groundwater flow, depending on the elevation of the well screen.

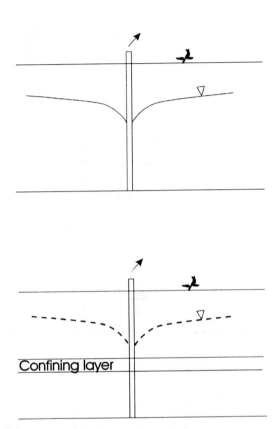

Figure 5.2. Cones of depression in unconfined (top) and confined (bottom) aquifers.

Steady Flow to Wells

For a confined aquifer under steady conditions, radial flow to a well is described by the Thiem (1906) equation (Figure 5.3):

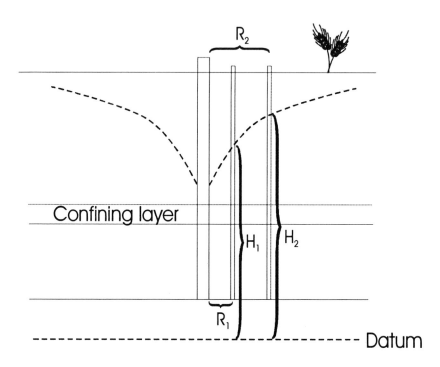

Figure 5.3. Pumping test parameters for steady, confined aquifer.

$$T = \frac{Q}{2\pi(H_2 - H_1)} \ln \frac{R_2}{R_1} \tag{5.1}$$

where Q is the pumping rate, H_2 is the hydraulic head at a distance R_2, and H_1 is the hydraulic head at a distance R_1. The distances R_1 and R_2 are measured horizontally from the pumping well, and the subscript 1 denotes the closer observation well. Any datum can be used for the hydraulic head values – the head difference is independent of the datum.

Example 5.1: Calculate the transmissivity of an aquifer for which the following data were obtained during a pumping test. The aquifer is confined, and steady conditions prevail.

pumping discharge = 220 gal/min
distance to observation well 1 = 26 ft
distance to observation well 2 = 73 ft
hydraulic head at observation well 1 = 29.34 ft
hydraulic head at observation well 2 = 32.56 ft

$$T = \frac{Q}{2\pi(H_2 - H_1)} \ln\frac{R_2}{R_1} = \frac{220\,\text{gal}}{\text{min}} \times \frac{1\,\text{ft}^3}{7.48\,\text{gal}} \times \frac{1}{2} \times \frac{1}{\pi} \times \frac{1}{3.22\,\text{ft}} \ln\left(\frac{73\,\text{ft}}{26\,\text{ft}}\right)$$
$$= 1.5\,\text{ft}^2/\text{min}$$

For an unconfined aquifer, a slightly different equation applies, employing K instead of T. In this case, the datum must coincide with the base of the aquifer (Figure 5.4).

$$K = \frac{Q}{\pi(H_2^2 - H_1^2)} \ln\frac{R_2}{R_1} \tag{5.2}$$

Equations 5.1 and 5.2 can be applied in a direct fashion, to predict H_1 or H_2 given values for the other variables. An inverse application would involve a pumping test, during which stabilized water levels in monitoring wells yield H_1, H_2, R_1, and R_2. From those values, an estimate of K or T can by made.

Example 5.2: A well in an unconfined aquifer is pumped at 43 gal/min. At steady conditions, the hydraulic head (meas-

ured from the horizontal base of the aquifer) is 14.7 ft at an
observation well located 19 ft from the pumping well. The
aquifer has a hydraulic conductivity of 11 ft/day. Estimate
the hydraulic head at an observation well located 150 ft from
the pumping well.

$$K = \frac{Q}{\pi(H_2^2 - H_1^2)}\ln\left(\frac{R_2}{R_1}\right) = \frac{43\,\text{gal}}{\text{min}}\times\frac{1{,}440\,\text{min}}{1\,\text{day}}\times\frac{1\,\text{ft}^3}{7.48\,\text{gal}}\times\frac{1}{\pi}\times$$

$$\frac{1}{(H_2^2 - 216.09\,\text{ft}^2)}\ln\left(\frac{150\,\text{ft}}{19\,\text{ft}}\right) = \frac{5{,}447\,\text{ft}^3/\text{d}}{(H_2^2 - 216.09\,\text{ft}^2)} \Rightarrow$$

$$11\frac{\text{ft}}{\text{d}}(H_2^2 - 216.09\,\text{ft}^2) = 5{,}447\,\frac{\text{ft}^3}{\text{d}} \Rightarrow H_2 = 26.7\,\text{ft}$$

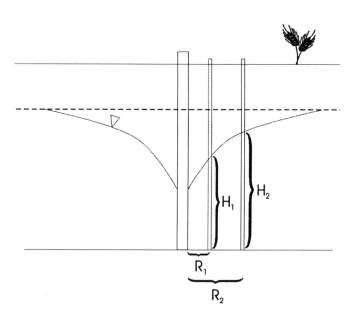

Figure 5.4. Pumping test parameters for steady, unconfined aquifer.

Transient Flow to Wells

Transient flow means that at any given point in an aquifer, the hydraulic head value is changing with time. Such a condition prevails during the early stages of pumping, before the cone of depression has stabilized. Water levels drop during a transient test, because water discharged by the pumping well is coming in part from the aquifer's storage. In contrast, during steady conditions, aquifer storage is not depleted – the discharged water is coming from groundwater flowing through the aquifer.

The Theis (1935) method can be used to measure the transmissivity and storage coefficient of a confined aquifer under transient flow conditions. Alternatively, given estimates of the transmissivity and storage coefficient, the equation can be applied to predict the effects of pumping a confined aquifer.

The equations used to relate drawdown in the observation well to the transmissivity and storage coefficient are:

$$s = \frac{Q}{4\pi T} W(u) \tag{5.3}$$

$$u = \frac{r^2 S}{4Tt} \tag{5.4}$$

where s is drawdown in the observation well, Q is discharge from the pumping well, T is aquifer transmissivity, $W(u)$ is the well function, r is the distance between the pumping well and observation well, S is the storage coefficient, and t is time since pumping began. Values of $W(u)$ corresponding to various values of $1/u$ are listed in Table 5.1. Interpolation can be used to estimate numbers that fall between the values in Table 5.1. More elaborate equations have been derived for transient flow in unconfined aquifers (Neuman, 1975), and for other complex conditions.

Table 5.1. Values of $W(u)$ versus $1/u$ (Heath, 1983).

1/u	10	7.69	5.88	5.00	4.00	3.33	2.86	2.5	2.22
0.1	0.219	0.135	0.075	0.049	0.025	0.013	0.007	0.004	0.002
1	1.82	1.59	1.36	1.22	1.04	0.91	0.79	0.70	0.63
10	4.04	3.78	3.51	3.35	3.14	2.96	2.81	2.68	2.57
10E2	6.33	6.07	5.80	5.64	5.42	5.23	5.08	4.95	4.83
10E3	8.63	8.37	8.10	7.94	7.72	7.53	7.38	7.25	7.13
10E4	10.94	10.67	10.41	10.24	10.02	9.84	9.68	9.55	9.43
10E5	13.24	12.98	12.71	12.55	12.32	12.14	11.99	11.85	11.73
10E6	15.54	15.28	15.01	14.85	14.62	14.44	14.29	14.15	14.04
10E7	17.84	17.58	17.31	17.15	16.93	16.74	16.59	16.46	16.34
10E8	20.15	19.88	19.62	19.45	19.23	19.05	18.89	18.76	18.64
10E9	22.45	22.19	21.92	21.76	21.53	21.35	21.20	21.06	20.94
10E10	24.75	24.49	24.22	24.06	23.83	23.65	23.50	23.36	23.25
10E11	27.05	26.79	26.52	26.36	26.14	25.96	25.80	25.67	25.55
10E12	29.36	20.09	28.83	28.66	28.44	28.26	28.10	27.97	27.85
10E13	31.66	31.40	31.13	30.97	30.74	30.56	30.41	30.27	30.15
10E14	33.96	33.70	33.43	33.27	33.05	32.86	32.71	32.58	32.46

1/u	2.00	1.67	1.43	1.25	1.11
0.1	0.001	0.000	0.000	0.000	0.000
1	0.56	0.45	0.37	0.31	0.26
10	2.47	2.30	2.15	2.03	1.92
10E2	4.73	4.54	4.39	4.26	4.14
10E3	7.02	6.84	6.69	6.55	6.44
10E4	9.33	9.14	8.99	8.86	8.74
10E5	11.63	11.45	11.29	11.16	11.04
10E6	13.93	13.75	13.60	13.46	13.34
10E7	16.23	16.05	15.90	15.76	15.65
10E8	18.54	18.35	18.20	18.07	17.95
10E9	20.84	20.66	20.50	20.37	20.25
10E10	23.14	22.96	22.81	22.67	22.55
10E11	25.44	25.26	25.11	24.97	24.86
10E12	27.75	27.56	27.41	27.28	27.16
10E13	30.05	29.87	29.71	29.58	29.46
10E14	32.35	32.17	32.02	31.88	31.76

Example: When $1/u = 3.33 \times 10^2$, $W(u) = 5.23$.

Example 5.3: Predict the drawdown 144 m from a pumping well after 7.3 hours, given a storage coefficient of 0.001, transmissivity of 100 m²/day, and discharge of 100 gal/min.

$$u = \frac{r^2 S}{4Tt} = (144 \text{ m})^2 \times 0.001 \times \frac{1}{4} \times \frac{1 \text{ d}}{100 \text{ m}^2} \times \frac{1}{7.3 \text{ h}} \times \frac{24 \text{ h}}{1 \text{ d}} =$$
$$0.170 \Rightarrow W(u) = 1.36$$

$$s = \frac{Q}{4\pi T} W(u) = \frac{100 \text{ gal}}{\text{min}} \times \frac{1,440 \text{ min}}{1 \text{ d}} \times \frac{1 \text{ ft}^3}{7.48 \text{ gal}} \times \left(\frac{1 \text{ m}}{3.28 \text{ ft}}\right)^3 \times$$
$$\frac{1}{4} \times \frac{1}{\pi} \times \frac{1 \text{ d}}{100 \text{ m}^2} \times 1.36 = 0.59 \text{ m}$$

An inverse application of the Theis method involves plotting the time-drawdown data recorded at an observation well on logarithmic paper. The data can be analyzed using manual curve matching or computer software. Curve matching entails the following steps. (1) Overlay a Theis type curve (Figure 5.5) on the time-drawdown data. Ensure that the type curve and data are plotted on identical scales (for example, if one log cycle equals one inch on the type curve, the same should hold true for the data). (2) Move the type curve, keeping its axes parallel to those of the field data, until it closely lines up with the field data. (3) Choose any point in the field of overlap between the two plots. Identify the values of $W(u)$, $1/u$, t, and s that correspond to this point. (4) Use these values with Equations (5.3) and (5.4) to solve for T and S.

Example 5.4: Estimate the transmissivity and storage coefficient of a confined aquifer for which the following pumping test data were obtained.

match point: $W(u) = 1$; $1/u = 1$; $s = 3.69$ m; $t = 2.31$ h

distance to observation well = 38 m

pumping rate = 114 m³/d

$$s = \frac{Q}{4\pi T} W(u) \Rightarrow T = \frac{QW(u)}{4\pi s} = \frac{114\,\text{m}^3}{\text{d}} \times 1 \times \frac{1}{4} \times \frac{1}{\pi} \times \frac{1}{3.69\,\text{m}} =$$

$$2.46\,\text{m}^2/\text{d}$$

$$u = \frac{r^2 S}{4Tt} \Rightarrow S = \frac{4Ttu}{r^2} = 4 \times \frac{2.46\,\text{m}^2}{\text{d}} \times 2.31\,\text{h} \times \frac{1\,\text{d}}{24\,\text{h}} \times 1 \times$$

$$\frac{1}{(38\,\text{m})^2} = 0.0007$$

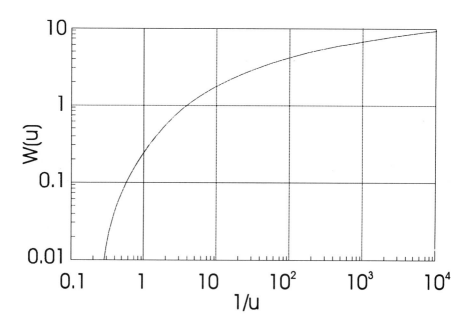

Figure 5.5. Theis type curve (modified from Heath, 1983).

Pumping Tests

Pumping tests are used to apply the preceding equations in an inverse manner. When conducted properly, they are among the most reliable methods of determining the transmissivity and storage coefficients of aquifers (EPA, 1991b). However, pumping tests require a greater degree of activity and expense than alternative methods and are not always justified for all levels of investigation. These tests involve pumping a well at a constant rate and measuring drawdown in nearby wells (Figure 5.6).

Before starting the test, water levels should be carefully monitored in the pumping and observation wells. The purpose of this preliminary monitoring, called a trend test, is to determine how water levels are fluctuating over time under pre-pumping conditions. For example, water level fluctuations can be caused by changes in barometric pressure or recharge from precipitation. The pre-pumping water level trend must be used to correct drawdown values measured during the pumping test. The corrected values should reflect only the effects of the pumping well.

During a pumping test, all depth-to-water measurements should be made from the surveyed reference points. The rate of drawdown is normally rapid at first, and then slower as time progresses. Therefore, readings should be taken more frequently at first, for example, every 30 seconds. For a confined aquifer, a pumping test may last 24 hours or less (Driscoll, 1986). Pumping tests in unconfined aquifers normally require more time, to account for delayed gravity drainage.

Observation wells should be located close enough to the pumping well to be impacted during the test, and screened in the same aquifer as the pumping well. Depth-to-water measurements in the observation wells and pumping well should be made on the same time schedule. Ideally, both the pumping and observation wells should fully penetrate the aquifer. However, it is more critical that the pumping well fully penetrate the aquifer to induce horizontal flow.

Site ID:			Distance from Pumped Well:	
Location:			Logger:	
Test Start			Test End	
Date:			Date:	
Time:			Time:	
Static Water Level:			Static Water Level:	
Average Pumping Rate:				
Measurement Method:				
Comments:				
Pump Test Elapsed Time	Pump Test Depth to Water	Pumping Rate	Recovery Test Elapsed Time	Recovery Test Depth to Water
0.00			0.00	

Figure 5.6. Pump/recovery test data sheet.

Site ID:			Date:	
Location:			Logger:	
Pump Test Elapsed Time	Pump Test Depth to Water	Pumping Rate	Recovery Test Elapsed Time	Recovery Test Depth to Water

Figure 5.6 (continued). Pump/recovery test data sheet.

Given horizontal flow, the depth at which an observation well is screened should not affect its hydraulic head measurement. There would be no vertical gradient, and thus no hydraulic head variability along a vertical transect. However, neither the pumping well nor the observation wells should go dry during a pumping test.

The pump and accessories must be reliable, capable of drawing water at a constant rate through the duration of the test. If a pump fails, the data may be insufficient to obtain reliable estimates of aquifer properties. Throughout the test, discharge must be measured from the pump, and the flow rate must be adjusted to maintain a constant rate. There must also be a means for conveying water into a holding tank or away from the test site. This is especially important for shallow unconfined aquifers that could be recharged by discharge from the pumping well. Such recharge would interfere with the interpretation of the test results.

After the end of a pumping test, the pump is shut off and recovery measurements are made in the test wells. In theory, the water levels will recover at the same rate they fall. Recovery data can be used to derive an additional estimate of aquifer transmissivity, which may be better than the first estimate. In some cases, there are uncontrolled variations in the pumping rate during the pumping test which affect the drawdown. However, such variations do not affect the recovery rate. The flow rate used for the recovery data is the mean discharge for the entire pumping period (Todd, 1980). The differences between water levels at the end of pumping and at various times since pumping stopped are plotted as a function of the time since pumping stopped.

Ideally, pumping tests should involve at least one observation well. Two wells are required for the steady flow equations. In a transient test, using only a pumping well can be problematic because drawdown in the well will likely exceed that in the adjacent aquifer. The difference between drawdown in the well and aquifer is the well loss. The efficiency of a well is defined as the ratio of the drawdown outside the well to the drawdown inside the well (Figure 5.7). (Observation wells are required to estimate drawdown in the aquifer.) In practice, a 70 to 80% efficiency is considered good.

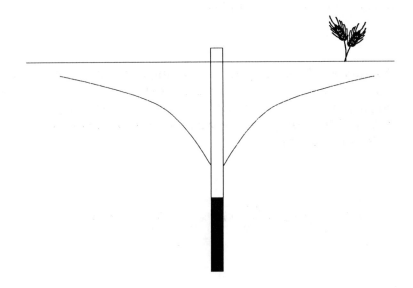

Figure 5.7. Drawdown inside and outside pumping well.

Heath (1983) reports an approximate solution for transmissivity in an aquifer test involving only a pumping well. It assumes that well loss increases total drawdown in the pumping well, but does not affect the rate of change in the drawdown with time. The solution is

$$T = \frac{2.3Q}{4\pi\Delta s} \qquad\qquad (5.5)$$

where Δs is the drawdown that occurs over one log cycle on a semilogarithmic plot of time (log scale) and drawdown (arithmetic scale). A single-well pumping test does not yield information on the storage coefficient of an aquifer.

Example 5.5: A single-well pumping test was conducted in a confined aquifer. Time-drawdown data were plotted on semi-logarithmic paper, and the drawdown over one log cycle was 3.7 m. The pumping rate was 24 m³/day. Estimate the aquifer's transmissivity.

$$T = \frac{2.3Q}{4\pi\Delta s} = 2.3 \times \frac{24\,\text{m}^3}{\text{d}} \times \frac{1}{4} \times \frac{1}{\pi} \times \frac{1}{3.7\,\text{m}} = 1.2\,\text{m}^2/\text{d}$$

Another parameter that can be obtained from a single-well test is the specific capacity of the pumping well. Specific capacity is a basic measure of the performance of a well, with higher values signifying a greater yield capability. The specific capacity of a pumping well is defined as

$$S_c = \frac{Q}{s} \qquad\qquad (5.6)$$

where s is the stabilized drawdown in the well.

Example 5.6: Calculate the specific capacity of a well given a discharge of 150 gal/min and a stabilized drawdown of 10 ft.

$$S_c = \frac{Q}{s} = \frac{150\,\text{gal}}{\text{min}} \times \frac{1}{10\,\text{ft}} = 15\,\text{gal/min/ft}$$

Slug Tests

Slug tests are an alternative to pumping tests for solving the inverse well hydraulics problem. Compared with pumping tests, slug tests can be performed quickly and at a low cost, because pumping and observation wells are not required. In a slug test, a known volume (slug) of water is suddenly added to or removed from a well, and the decline or recovery of the water level is measured at closely spaced time intervals.

Hvorslev (1951) devised a method for computing the hydraulic conductivity of a formation from slug test data. In that method, the unrecovered head difference, normalized to the initial head difference (Figure 5.8), is plotted on a logarithmic scale against time, which is plotted on a standard arithmetic scale (Figures 5.9 and 5.10). A line is fitted to the data, and the time at which the normalized unrecovered head difference = 0.37 is determined. This time, T_0, is then used with the following equation to estimate hydraulic conductivity, K.

$$K = \frac{r_p^2 \ln(L_i/r_i)}{2L_i T_0} \tag{5.7}$$

where r_p is the radius of the piezometer, L_i is the length of the intake (screened interval and sand filter), and r_i is the radius of the intake (including the sand filter). The analysis assumes an isotropic, homogeneous medium.

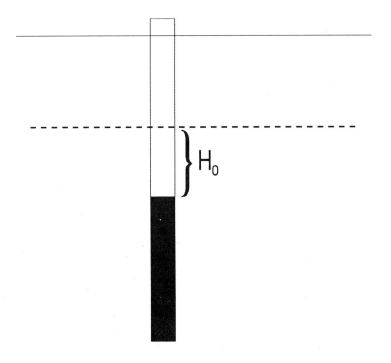

Figure 5.8. Initial head difference during slug test.

Example 5.7: Use the following slug test data to calculate a hydraulic conductivity value.

r_p = 0.083 ft
L_i = 3 ft
r_i = 0.3 ft
T_0 = 0.59 days

Date:			
Site ID:		Slug Volume:	
Location ID:		Logger:	
Test Method: _ Falling Head _ Rising Head			
Comments:			
Time (Begin Test 1):		Time (Begin Test 2):	
Time (End Test 1):		Time (End Test 2):	
Elapsed Time	Depth to Water	Elapsed Time	Depth to Water

Figure 5.9. Slug test data form.

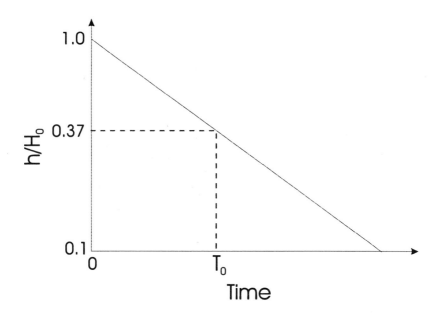

Figure 5.10. Plot of h/H_0 (logarithmic axis) versus time (arithmetic axis).

$$K = \frac{r_p^2 \ln(L_i / r_i)}{2L_i T_0} = (0.083 \text{ ft})^2 \ln\left(\frac{3 \text{ ft}}{0.3 \text{ ft}}\right) \times \frac{1}{2} \times \frac{1}{3 \text{ ft}} \times \frac{1}{0.59 \text{ d}} =$$

0.0045 ft/d

The hydraulic conductivity value obtained from a slug test applies only to the aquifer deposits in the vicinity of the well screen. In contrast, a pumping test gives a hydraulic conductivity value indicative of a broader portion of the aquifer.

Key Terms

cone of depression, direct approach, drawdown, inverse approach, pumping test, pumping well, slug test, specific capacity, steady flow, transient flow, trend test

Problems

1. Calculate a hydraulic conductivity value using the following slug test data. The data were collected during a rising-head test, with a static depth to water of 4.05 ft.

    r_p = 1 in
    r_i = 1 in
    L_i = 10 ft

Time (seconds)	Depth to Water (ft)
0	5.00
13	4.96
19	4.93
25	4.91
29	4.84
41	4.75
52	4.62
70	4.54
99	4.46
108	4.37
132	4.27
151	4.05

2. Estimate the transmissivity of a confined aquifer under steady flow given the following data:

    Q = 220 gal/min
    H_1 = 28.75 ft
    R_1 = 25 ft

H_2 = 32.01 ft
R_2 = 68 ft

3. Calculate the hydraulic conductivity of an unconfined aquifer using the following data:

    Q = 145 gal/min
    H_1 = 6.5 m
    H_2 = 8.9 m
    R_1 = 4.8 m
    R_2 = 73 m

4. A well is located in a confined aquifer with a hydraulic conductivity of 14 m/d, storage coefficient of 0.006, and saturated thickness of 19 m. The well pumps 2,650 m³/d. Predict the drawdown 7.0 m from the well after one day of pumping.

5. Plot the following pumping test data on logarithmic paper. The test was conducted in a confined aquifer, at a pumping rate of 26.8 ft³/s. Estimate the aquifer's transmissivity and storage coefficient. The observation well is 250 ft from the pumping well.

Time (min)	Drawdown at Observation Well (ft)
0	0.00
1	0.66
1.5	0.87
2	0.99
2.5	1.11
3	1.21
4	1.36
5	1.49
6	1.59
8	1.75

10	1.86
12	1.97
14	2.08
18	2.20
24	2.36
30	2.49
40	2.65
50	2.78
60	2.88
80	3.04
100	3.16
120	3.28
150	3.42
180	3.51
210	3.61
240	3.67

6. The following time-drawdown data were collected at a pumping well in a confined aquifer. Plot the data on semi-logarithmic paper, and estimate the aquifer's transmissivity. The pumping rate was 15 m^3/d.

Time (min)	Drawdown (m)
1	0.15
2	0.23
3	0.28
4	0.34
5	0.35
6	0.37
7	0.39
8	0.41
9	0.42
10	0.45

20	0.54
30	0.65
40	0.67
50	0.71
60	0.72
70	0.74
80	0.77
90	0.80
100	0.81
200	0.90
400	0.99
800	1.09
1000	1.11

7. Calculate the specific capacity of a well, given a pumping rate
    of 60 gal/min and a stabilized drawdown of 14 ft.

CHAPTER 6

Groundwater Quality and Solute Transport

Solutes

A solute is any substance dissolved in solution. Solutes in groundwater can occur naturally, or they can be introduced by various sources of pollution. The natural composition of groundwater is derived from many different sources, including gases from the atmosphere, weathering and erosion of rocks and soil, and chemical reactions beneath the land surface (Hem, 1970; Heath, 1983).

As water infiltrates a recharge area, solute concentrations are typically low. However, water quality changes along a flow path – solute concentrations increase with subsurface residence time. Groundwater in a local discharge area (typically associated with flow paths less than 100 m) will tend to be less mineralized than water issuing from a regional discharge zone (typically associated with flow paths exceeding 1 km) (Figure 6.1).

Groundwater also tends to evolve from Ca^{+2} (calcium ion) and HCO_3^{-1} (bicarbonate ion) rich in recharge areas to Na^{+1} (sodium ion) and Cl^{-1} (chloride ion) rich toward discharge areas. Minerals such as calcite and feldspar are dissolved by low pH rainwater in the recharge area, releasing calcium and bicarbonate ions to water. Groundwater transports calcium ions, which eventually exchange for sodium on clay mineral surfaces. Major controls on this hydrogeochemical evolution of groundwater are aquifer mineralogy, water rock interactions, flow velocity, distance along flow paths, residence time, and mixing.

Common groundwater solutes, usually present in concentrations above 1 mg/L, include calcium (Ca^{+2}), sodium (Na^{+1}), magnesium (Mg^{+2}), potassium (K^{+1}), bicarbonate (HCO_3^{-1}), sulfate (SO_4^{-2}), chloride (Cl^{-1}), and silica (SiO_2) (Fetter, 1994).

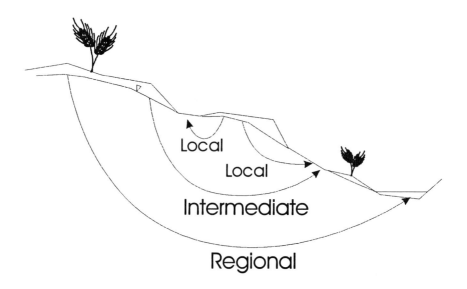

Figure 6.1. Local, intermediate, and regional flow systems.

Secondary ions in groundwater include iron (Fe^{+2}), nitrate (NO_3^{-1}), fluoride (F^{-1}), and boron (B^{+3}). The major gases dissolved in groundwater are oxygen (O_2) and carbon dioxide (CO_2). Nitrogen gas (N_2) is also present. Minor gases include hydrogen sulfide (H_2S) and methane (CH_4). Hydrogen sulfide and methane are produced when organic matter decays under anaerobic (oxygen-deficient) conditions.

The concentration of a solute in a sample of groundwater is usually expressed as either (1) the weight of the solute divided by the weight of water in the sample, or (2) the weight of the solute divided by the volume of water in the sample. Commonly used concentration units are parts per million (ppm) and milligrams per liter (mg/L). These units are approximately equal, provided that the groundwater's density is close to 1.00 g/cm³. For salty water, the density is higher than 1.00 g/cm³, and ppm and mg/L are not equivalent units.

> **Example 6.1:** The density of a solution is 1.14 g/cm³ (1,140,000 mg/L). If the concentration of chloride in the solution is 345 mg/L, what is its concentration in ppm?

$$\frac{345 \text{ mg}}{1,140,000 \text{ mg}} = \frac{X \text{ mg}}{1,000,000 \text{ mg}} \Rightarrow X = 303 \text{ mg} \Rightarrow 303 \text{ ppm}$$

Sources of Contamination

Groundwater can be contaminated by a variety of different sources. Some of the most common are landfill leachate, chemicals in underground tanks, septic tank effluent, industrial waste lagoons, chemicals disposed through injection wells, oil field brine, crop pesticides and fertilizers, animal feedlots, and saltwater intrusion.

Leachate forms when percolating rainwater dissolves certain constituents of solid waste. This liquid can seep out the bottom or sides of a landfill. Many modern landfills are equipped with clay or synthetic caps, liners, and leachate collection systems. Leachate collection systems consist of perforated pipe in a gravel blanket, above the liner. Caps exclude infiltration of rainwater to inhibit leachate formation. A problem with this strategy is that water accelerates waste decomposition. Without water, solid waste can remain potentially hazardous for hundreds of years (Freeze and Cherry, 1979).

Septic systems discharge household wastewater to the subsurface (Figure 6.2). Bacteria, viruses, nitrate, and phosphate are principal pollutants of household wastewater. In a conventional septic system, wastewater flows into an underground tank, where the solids settle out and are broken down by anaerobic bacteria. The liquid fraction moves through the tank to a series of perforated pipes. These pipes release the water to the environment, where it undergoes aerobic breakdown and filtration. In theory (but not always in practice), any wastewater that reaches the water table is clean.

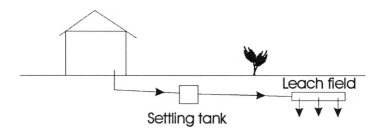

Figure 6.2. Conventional septic system.

Unfortunately, conventional septic systems often fail for numerous reasons, including clogged tanks and pipes, broken pipes, household chemicals that kill bacteria in the tank, and leach fields that are too close to the water table. For these reasons, aerobic spray systems are gaining popularity among builders in rural areas. These systems route the wastewater through a series of tanks, where the waste settles and is broken down aerobically. Air is pumped in to promote aerobic decay, a more efficient process than anaerobic decay. Another tank contains chlorine to kill the bacteria. After that, the water is sprayed onto the ground. Aerobic spray systems pose less threat to groundwater, but require careful maintenance – tanks should be pumped periodically and chlorine tablets replaced.

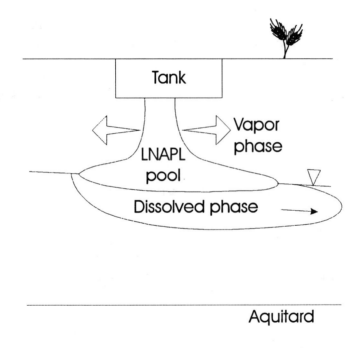

Figure 6.3. LNAPL spill.

Chemical spills often introduce nonaqueous-phase liquids (NAPLs) to the subsurface. These liquids can be lighter (LNAPLs) or denser (DNAPLs) than water. LNAPLs include hydrocarbon fuels such as gasoline, heating oil, kerosene, jet fuel, and aviation gas. DNAPLs include chlorinated hydrocarbons, such as trichloroethyl-

ene, carbon tetrachloride, chlorophenols, chlorobenzenes, tetra-
chloroethylene, and polychlorinated biphenyls (PCBs) (EPA, 1991b).

Spilled LNAPLs and DNAPLs form plumes of contamination in
the dissolved, nonaqueous, and vapor phases. LNAPLs "float" on
the capillary fringe (Figure 6.3), whereas DNAPLs migrate deep into
aquifers (Figure 6.4). Once a DNAPL reaches the bottom of an aquifer, it may move in the direction that the surface of the underlying aquitard is sloping (in response to gravity). In this way, a
DNAPL can actually move in the opposite direction that the
groundwater is flowing.

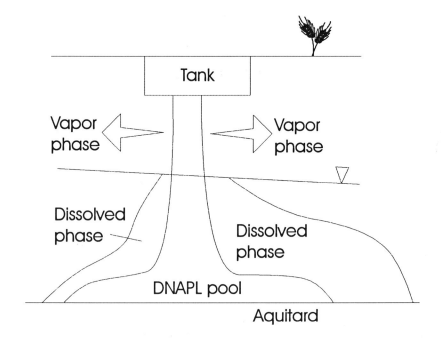

Figure 6.4. DNAPL spill.

Nitrate is a frequent groundwater contaminant in rural settings. Rural environments often contain high concentrations of household septic systems and agricultural activities that produce nitrogen. In aerated soils, nitrogen is converted to nitrate, which is mobile in groundwater and can be harmful to humans. Levels of nitrate in groundwater beneath agricultural areas are usually spatially heterogeneous (Figure 6.5). Clusters of high nitrate levels may reflect spatial variations in such factors as depth to groundwater, well depth, vadose-zone permeability, fertilizer application, and concentration of farm animals. High nitrate levels are normally associated with shallow water tables, shallow wells, permeable vadose zones, high rates of fertilizer application, and high concentrations of farm animals.

Saltwater intrusion is a common problem in coastal cities that pump groundwater. In coastal settings, underground freshwater resides shoreward and above underground saline water (Figure 6.6). Pumping can draw the freshwater/saltwater interface into a well, contaminating the water with high concentrations of dissolved solids.

According to the Ghyben-Herzberg principle, a 1 m drop in the water table of an unconfined, coastal aquifer will induce a 40 m rise of the freshwater/saltwater interface (any other units can be substituted for meters). This explains why saltwater intrusion is a common groundwater problem in coastal areas.

$$z = \frac{\rho}{(\rho_t - \rho)} H \approx 40H \tag{6.1}$$

where z is the vertical distance between sea level and the interface, H is the vertical distance between sea level and the water level in the well, ρ is freshwater density, and ρ_t is saltwater density.

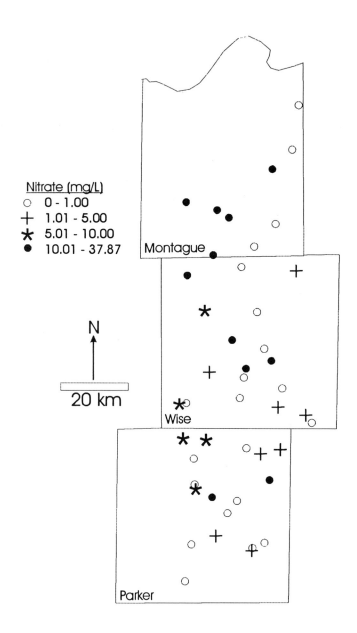

Figure 6.5. Nitrate levels in groundwater in a three-county area of north-central Texas (Hudak and Blanchard, 1997). Reprinted with permission from Elsevier Science.

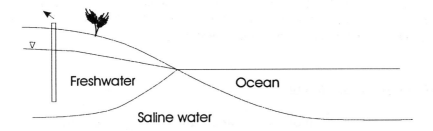

Figure 6.6. Coastal pumping well.

Example 6.2: Calculate z given that the distance H between sea level and the water table is 18 m at a coastal well.

$$z = 40H = 40 \times 18 \text{ m} = 720 \text{ m}$$

Water Quality Standards

To determine whether or not an aquifer is contaminated, hydrogeologists collect groundwater samples and have them chemically analyzed. If the concentrations of chemicals in groundwater exceed maximum contaminant levels, remedial action may be necessary. Maximum contaminant levels (MCLs) are enforceable standards established by the U.S. Environmental Protection Agency. States can set maximum contaminant levels that are stricter than the federal standards, but not less strict. The EPA also sets suggested maximum contaminant levels (SMCLs) for many chemicals. These are recommended but non-enforceable levels. Table 6.1 shows MCLs for some common chemicals.

Groundwater Sampling

A developed monitoring well can be used to extract ground-water samples from an aquifer. Prior to obtaining a sample, the well must be evacuated to remove stagnant water, which may not be representative of groundwater in the aquifer. Removing three times the volume of water standing in the well casing performs well evacuation.

Table 6.1. Water quality standards for selected chemicals (EPA, 1994b).

Contaminant	MCL (mg/L)
Inorganics	
Arsenic	0.05
Barium	2
Cadmium	0.005
Chromium (total)	0.1
Cyanide	0.2
Fluoride	4.0
Mercury	0.002
Nickel	0.1
Nitrate (as Nitrogen)	10
Nitrite (as Nitrogen)	1
Selenium	0.05
Uranium	0.02

Table 6.1. (continued)

Contaminant	MCL (mg/L)
Volatile Organics	
1,2-Dichloroethane	0.005
1,1,1-Trichloroethane	0.2
Benzene	0.005
Carbon Tetrachloride	0.005
Trichloroethylene	0.005
Vinyl Chloride	0.002
Organics	
Alachlor	0.002
Aldicarb	0.003
Atrazine	0.003
Carbofuran	0.04
Chlordane	0.002
Chlorobenzene	0.1
Dinoseb	0.007
Diquat	0.02
Dioxin	0.00000003
Endrin	0.002
Ethylbenzene	0.7
Ethylene dibromide	0.00005
Glyphosate	0.7
Heptachlor	0.0004
Lindane	0.0002
PCBs	0.0005
Toluene	1
Toxaphene	0.003
Xylenes (total)	10

Example 6.3: A monitoring well contains a 10-ft water column prior to sampling. The well's diameter is 4 inches. It will be evacuated with a 3-ft long bailer having a diameter of 2 inches. How many bailer volumes are required to evacuate three times the initial volume in the well?

$$\text{initial volume in well} = \pi \times \text{radius}^2 \times \text{height} = 0.87 \, \text{ft}^3$$
$$\text{volume of bailer} = \pi \times \text{radius}^2 \times \text{height} = 0.065 \, \text{ft}^3$$
$$\text{bailer volumes required} = 3 \times 0.87 \, \text{ft}^3 / 0.065 \, \text{ft}^3 = 40.2$$

Evacuation can be performed with a pump or bailing device. The water level in an evacuated well should be allowed to recover to the original level before sampling.

The device used to collect a groundwater sample must not alter the chemistry of the sample as it is brought to the surface. This could occur if the materials from which the device is constructed leach compounds into the sample or absorb compounds from the sample. Peristaltic pumps, electric submersible pumps, bailers and bladder pumps are commonly used to obtain groundwater samples from wells.

Peristaltic pumps employ a rotating disk that massages a flexible plastic tube. The massaging action creates a suction that draws water upward from the well. In contrast, bladder pumps use pulses of air to displace a water sample through tubing to the land surface. Electric submersible pumps are attached to a hose that is lowered down the well. A propeller device draws water into the pump. It then travels through the hosing to the land surface.

Bailers are cylindrical, constructed of PVC, Teflon, or stainless steel. They are lowered on a rope down the well casing. A check valve at the bottom of the bailer allows water to enter, but seals under the weight of a water column when it is lifted out of the well. Bailers are relatively inexpensive and easy to use, requiring no external power source. However, they can also be time-consuming and labor intensive. Another disadvantage is that the transfer of water to a sample container may alter the chemistry of the sample due to volatilization or aeration. It is also difficult to determine the exact location in the water column from which a bailed sample was collected. Pumps can be used to overcome some of the problems associated with bailers.

Samples should be collected as soon as possible after a well is purged. Ideally, the rate of sample collection should be approximately the same as the actual groundwater flow rate. The EPA suggests low sampling rates, approximately 0.1 liters per minute. Samples should be placed in clean, air-tight bottles. There should be no air space in samples used for volatile compounds or dissolved gases.

The type of analysis for which a sample is being collected determines the type of bottle, preservative, holding time, and filtering requirements. Preservatives can retard biological action, inhibit chemical reactions such as hydrolysis or oxidation, and prevent sorption. Generally, samples should be placed in a cooler maintained at 4°C. Ideally, they should be shipped to an analytical laboratory within 24 hours of sample collection.

Sample labels should include the identification number, name of the collector, date and time collected, place of collection, and parameters to be analyzed by the laboratory. A chain of custody procedure should be designed to allow the owner/operator of a site to reconstruct how and under what circumstances a sample was collected (Figure 6.7). The purpose of this procedure is to prevent misidentification of samples, prevent tampering with the samples during shipping and storage, allow easy identification of any tampering, and permit easy tracking of possession.

Hydrogeologists should develop a quality control program to ensure that the analytical results (reported by a laboratory) accurately express the actual concentrations of chemicals in groundwater. Spiked samples check laboratory accuracy. A spiked sample contains a known concentration of a particular solute. It is submitted to the lab along with the other samples. The lab report is checked to verify that the proper concentration is reported for the spiked sample.

Analytical precision is the ability of the laboratory to reproduce results. Split samples, used to check laboratory precision, can be made by mixing a larger sample and splitting it into two containers. Because the split samples are from the same source, they should yield identical concentrations when they are chemically analyzed.

Field blanks can be used to assess the quality of a field-sampling program. Purified water is taken to the field in a sealed container. The water is run through the sampling equipment, placed in a container, and sent to the lab for analysis. It should not contain any contaminants. The presence of one or more contaminants is an indication that the sampling equipment was not properly cleaned between the collection of samples. When sampling several wells during a field outing, a collector should start with the cleanest well and move toward progressively dirtier wells. This practice reduces the chance of cross-contamination.

Electrical Conductance and TDS

Some water quality indicators can be measured directly in the field. The conductivity of a solution is a measure of its ability to carry an electrical current. Typical units for electrical conductance are millisiemens (mS) and microsiemens (µS). Pure liquid water has a very low electrical conductance, less than a tenth of a microsiemen at 25°C (Hem, 1970). Charged ionic species make a solution conductive. Most substances dissolved in water dissociate into ions that can conduct an electrical current. The type and number of ions in the solution also affect conductance. In general, the larger the conductance, the more mineralized the water.

Sampler's Signature	Site	Date	Sample # to Sample #	Log ID # to Log ID #
Relinquished by:				Date & Time
Received by:				Date & Time
Relinquished by:				Date & Time
Received by:				Date & Time
Relinquished by:				Date & Time
Received by:				Date & Time
Relinquished by:				Date & Time
Received by:				Date & Time
Relinquished by:				Date & Time
Received by:				Date & Time

Figure 6.7. Chain-of-custody form.

Electrical conductance measurements can be used to estimate the concentration of total dissolved solids (TDS) in a water sample. However, it should be noted that natural water contains a variety of both ionic and uncharged chemical species. Therefore, conductance determinations cannot be used to obtain highly accurate estimates of TDS. To convert conductance to TDS, the following relation is used (Hem, 1970):

$$\text{TDS} = BC_m \tag{6.2}$$

where C_m is the conductance in microsiemens, TDS is expressed in milligrams per liter, and B is a conversion factor. For most groundwater, B is between 0.55 and 0.75, depending on the ionic composition of the solution.

A second, more accurate way to measure TDS is to evaporate a known volume of a sample at 105°C and weigh the residue. By summing the concentrations measured for individual solutes, a third estimate of TDS can be made. However, this third approach requires that every solute potentially present in the water is measured – an impractical and costly endeavor.

Example 6.4: Estimate the TDS of a sample having a conductance of 945 μS.

$$\text{TDS} = 0.65 \times 945 = 614 \text{ mg/L}$$

The concentration of dissolved minerals in groundwater is a general indication of its suitability for a particular use (Driscoll, 1986). The natural quality of groundwater varies substantially. TDS can range from less than 100 mg/L to more than 100,000 mg/L. High values characterize brines in many deep aquifers. Freshwater contains less than 1,000 mg/L (Table 6.2).

Barring the presence of contaminants, water containing less than 500 mg/L dissolved solids is generally satisfactory for domestic use and many industrial purposes. If total dissolved solids exceed 1,000 mg/L, the electrical conductivity of the water may cause serious electrolytic corrosion of metal well screens or pump components (Driscoll, 1986). An increase in TDS near a potential source of groundwater contamination, such as a landfill, may be an indication that a leak has occurred. More detailed sampling could define the problem.

Table 6.2. Total dissolved solids classifications (Hem, 1970).

Category	TDS (mg/L)
Fresh	< 1,000
Moderately Saline	3,000-10,000
Very Saline	10,000-35,000
Briny	> 35,000

Solute Transport

There are two main processes that affect the transport of solutes in groundwater: advection and hydrodynamic dispersion. Advection occurs when flowing groundwater carries along dissolved solutes. Hydrodynamic dispersion is the spreading of groundwater and its dissolved constituents by mechanical mixing and diffusion. Mechanical mixing is spreading caused by the aquifer matrix. Diffusion, which can take place in groundwater that is not flowing, is the movement of solutes from areas of high to low concentration. This latter process can be illustrated by placing a drop of dye in a beaker of stagnant water. The dye will spread even though the water is not flowing.

Mixing along a flow path is called longitudinal dispersion. Transverse (perpendicular) dispersion takes place both horizontally and vertically. At the pore scale, mechanical causes of longitudinal dispersion include (1) fluid moving faster in the center of pores than along the edges, (2) some fluid traveling along longer pathways than other fluid, and (3) fluid in larger pores traveling faster than fluid in smaller pores (Fetter, 1994). Flow paths bifurcating laterally cause transverse dispersion. On a larger scale, mechanical dispersion is also caused by heterogeneities in the aquifer, a process known as macro-dispersion.

Solutes will spread more in the direction of groundwater movement than in the perpendicular direction. Consequently, the magnitude of longitudinal dispersion exceeds that of transverse dispersion. The coefficients of longitudinal, transverse horizontal, and transverse vertical hydrodynamic dispersion are defined as:

$$D_L = \alpha_L v + D^*$$ (6.3)

$$D_{TH} = \alpha_{TH} v + D^*$$ (6.4)

$$D_{TV} = \alpha_{TV} v + D^*$$ (6.5)

where α is the dispersivity of the medium and $D*$ is the effective molecular diffusion coefficient. $D*$ ranges from 10^{-11} to 10^{-8} m²/s for most ions in groundwater (Fetter, 1994). Dispersivity quantifies the ability of a medium to induce spreading by mechanical means. Generally, diffusion is a significant component of Equations 6.3 to 6.5 only for low flow velocities. Dispersivity and the diffusion coefficient have units of length and length-squared/time, respectively.

Example 6.5: Calculate the mechanical mixing term of longitudinal hydrodynamic dispersion using the following data. If $D* = 5 \times 10^{-10}$ m²/s, is the system dominated by mechanical mixing or chemical diffusion?

groundwater velocity = 0.4 m/d
longitudinal dispersivity = 1.0 m

$$\alpha_L v = 1.0\,\text{m} \times \frac{0.4\,\text{m}}{\text{d}} = 0.4\,\text{m}^2/\text{d}$$

$$\frac{5 \times 10^{-10}\,\text{m}^2}{\text{s}} \times \frac{86{,}400\,\text{s}}{1\,\text{d}} = 4.3 \times 10^{-5}\,\text{m}^2/\text{d}$$

(dominated by mechanical mixing)

Dispersivity is a scale-dependent parameter, attaining higher values as the length of the transport domain increases. Lab estimates of dispersivity are routinely smaller than field estimates. In a critical review of dispersivity observations from 59 different field sites, Gelhar and others (1992) found that, for data of high reliability, longitudinal dispersivity ranged from 0.4 to 4 m. Transverse horizontal dispersivity was typically an order of magnitude less than longitudinal dispersivity, and transverse vertical dispersivity was an order of magnitude smaller than transverse horizontal dispersivity.

A tube filled with sand illustrates hydrodynamic dispersion. Distilled water runs through the tube at a constant rate. The chloride concentration of the influent is suddenly changed from zero to C_0 (for example, 1,000 mg/L). Chloride concentrations, C, are monitored at the outlet, and the ratio C/C_0 is plotted against time (Figure 6.8).

Early on, there are no chloride ions at the outlet, and C equals 0. Among the first group of chloride ions entering the tube, some arrive at the outlet ahead of others (due to hydrodynamic dispersion). As more chloride ions reach the outlet, C gradually increases to C_0. Had the solute been transported only by advection, the plotted curve would rise suddenly rather than gradually to C_0. The solute front "breaks through" the outlet when C/C_0 equals 0.5.

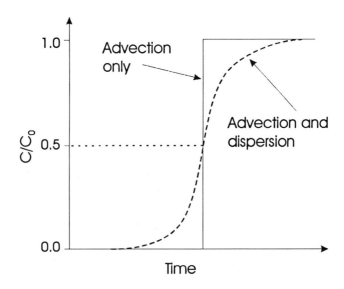

Figure 6.8. Breakthrough curve.

Solutes moving through groundwater may be conservative or reactive. Conservative solutes do not react with the aquifer medium or groundwater. Examples of conservative solutes are nitrate, chloride, and bromide. Reactive solutes have a tendency to adsorb to an aquifer medium, thereby moving from the liquid to solid phase. Many synthetic organic chemicals are reactive, as are many cations (positively charged ions). Clay is a good adsorbing medium. If a solute is reactive, it will travel at a slower rate than the groundwater.

The velocity at which a reactive solute travels relative to flowing groundwater depends on both the solute and aquifer material. A retardation factor, R_f, of 2 implies that the reactive solute travels at 1/2 the velocity of groundwater, a factor of 5 implies 1/5 the velocity of groundwater, and so forth. Higher R_f values indicate slower solute transport. R_f can be computed as:

$$R_f = 1 + \frac{\rho_b}{n} K_d \qquad (6.6)$$

where K_d (commonly mL/g or L/kg), measured in a laboratory, is the distribution coefficient for the solute with the aquifer. A high K_d value means that the solute has a strong adsorption tendency. For conservative solutes, K_d is equal to 0.

Example 6.6: Compute the relative velocity of a solute front, given that K_d = 10 mL/g, n = 0.30, and ρ_b = 1.83 g/cm³.

$$R_f = 1 + \frac{\rho_b}{n} K_d = 1 + \frac{1.83\,\text{g}}{\text{cm}^3} \times \frac{1}{0.30} \times \frac{10\,\text{mL}}{\text{g}} \times \frac{1\,\text{cm}^3}{1\,\text{mL}} = 62 \Rightarrow$$

relative velocity = 1/62

The distribution coefficient for a compound with a particular soil can be determined in a laboratory. An analyst fills jars containing soil with solutions of various contaminant concentrations. Several batches are created in this manner. The concentration of the contaminant in solution ($C_{solution}$) and the concentration in soil (C_{soil}) are measured for each batch. These data are plotted on a graph of C_{soil} versus $C_{solution}$, and a curve is fit through the points. The resulting curve (isotherm) can take on different shapes. In the simplest case, the curve is linear, and K_d is the slope of the line (Figure 6.9).

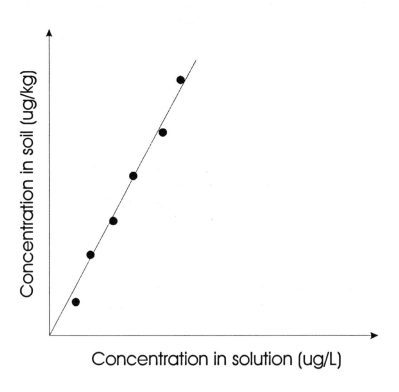

Figure 6.9. Data from laboratory batch experiment.

When a pollution source is leaking continuously over time, the highest contaminant concentrations in groundwater will typically occur beneath or near the source (Figure 6.10). As contaminated groundwater moves away from the source, it becomes diluted, and contaminant concentrations decrease. A continuously leaking source yields a plume that is connected to the source and extends downgradient (Figure 6.11). If the source stops leaking, the contaminant plume will move as a slug through the aquifer (Figure 6.12). It may be difficult to identify the source of contamination if a slug has moved a significant distance.

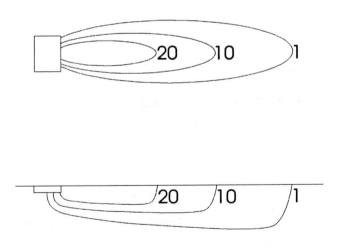

Figure 6.10. Contaminant concentration contours for continuous leaking source in map view (top) and side view (bottom).

In the field, the shape of a contaminant plume depends on several factors, including groundwater velocity, the hydraulic conductivity distribution of aquifer deposits, the shape of the water table, and the shape of the contaminant source. Figure 6.13 shows a contaminant plume near a landfill in Butler County, Ohio. The landfill occupies a glacial outwash (sand and gravel) aquifer bounded laterally and below by shale bedrock. Note that the shape of the plume bears some resemblance to the shape of the landfill, though it has grown in size due to dispersion. The plume has grown primarily in the downgradient direction, under the influence of the prevailing water table (Figure 4.8).

It is often necessary to predict how contaminants will move under ambient groundwater flow conditions. Such a prediction can be made by constructing flow lines originating from points along the perimeter of a contaminant plume. The area traversed by the flow lines defines a probable impact zone (Figure 6.14).

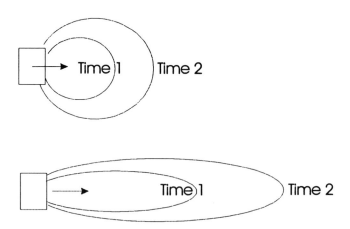

Figure 6.11. Contaminant plumes for continuous leaking source in slow (top) and fast (bottom) moving groundwater.

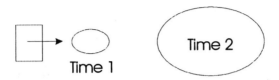

Figure 6.12. Contaminant slug.

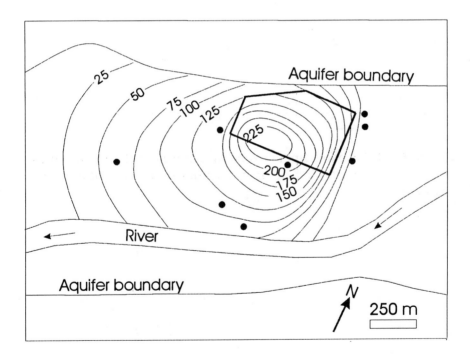

Figure 6.13. Contaminant plume (contours in mg/L) beneath solid-waste landfill in southwest Ohio; dots represent observation wells (Hudak and Loaiciga, 1991). Reprinted with permission from American Society of Civil Engineers.

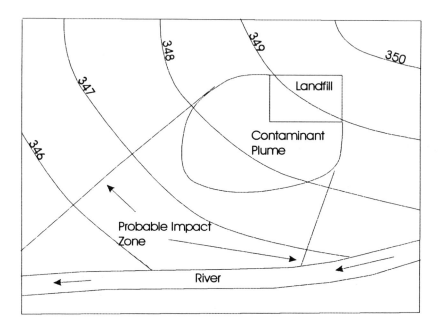

Figure 6.14. Probable impact zone for hypothetical landfill; numbers marking hydraulic head contour lines in meters above mean sea level.

Key Terms

adsorption, advection, chain of custody, concentration, conservative solute, diffusion, dispersivity, distribution coefficient, DNAPL, electrical conductance, field blank, Ghyben-Herzberg principle, hydrodynamic dispersion, leachate, LNAPL, maximum contaminant level, nitrate, reactive solute, retardation factor, saltwater intrusion, solute, spiked sample, split sample, total dissolved solids, well evacuation

Problems

1. Figure 6.15 shows five groundwater monitoring wells at a gasoline station in Fort Worth, Texas. Each well is labeled with a unique identifier. Concentrations of toluene measured in samples from the wells are listed below. (Toluene is a constituent of gasoline.) Make a copy of the page, and use the data to construct a toluene concentration contour map. A colored pencil should be used for the concentration contours to distinguish them from the water table contours on the figure. Draw contours of 1,000, 100, 10, and 2 parts per billion (ppb). (The 2-ppb contour defines the boundary of the plume.)

Well	Toluene Concentration (ppb)
MW-1	150
MW-2	<2
MW-3	1,900
MW-4	10
MW-5	8

What is the probable source of contamination? Shade the area that will likely be traversed by the contaminant plume if it is not contained or remediated.

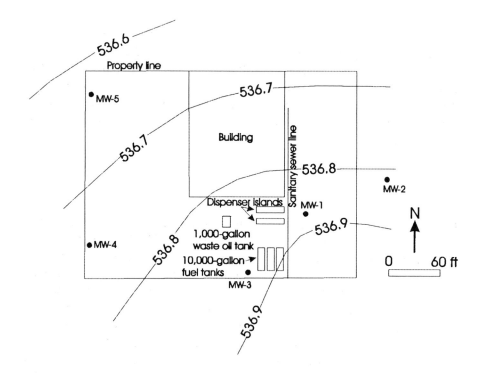

Figure 6.15. Site map for groundwater contamination problem; water table contours in feet above mean sea level.

2.   Estimate the concentration of total dissolved solids in a groundwater sample with an electrical conductance of 750 μS.

3.   A 150-mL sample of groundwater evaporated, leaving behind 37 mg of residue. Estimate the TDS of the sample.

4.   Distinguish between mechanical dispersion and chemical diffusion.

5.   At a coastal well, the distance from sea level to the freshwater/saltwater interface is 160 m. The bottom of the well is 15 m below sea level, and the water table is 4 m above sea level. How far can the water table be drawn down before saltwater enters the well?

6.   In what media would you expect molecular diffusion to dominate hydrodynamic dispersion?

7.   Calculate a retardation factor given the following data:

    distribution coefficient = 11 mL/g
    bulk density = 1.8 g/cm³
    porosity = 0.24

8.   What are the principal sources of groundwater contamination in your hometown? Explain.

9.   Calculate a K_d value given the following data (continued on next page) from a laboratory batch experiment. Plot the data on graph paper.

Concentration in Soil (mg/kg)	Concentration in Solution (mg/L)
40	0.11
71	0.25
159	0.48
240	0.73
282	0.81

CHAPTER 7

Managing Groundwater Pollution

After a plume of contamination has been identified and the properties of the aquifer have been carefully studied, hydrogeologists can devise a corrective action plan (CAP). Aquifer remediation is both time-consuming and expensive. Many problems require several years and cost hundreds of thousands of dollars.

The first step in restoring contaminated groundwater is to control the source. It must be prevented from releasing additional contaminants to the subsurface. Essentially, there are two approaches to controlling a contaminant source. Either the source is excavated and removed, or it is isolated in place. Excavation and removal may not be feasible if the source is laterally extensive and deep. Furthermore, excavating the source requires that there be a nearby facility willing to accept the excavated waste.

The alternative source control measure, waste isolation, involves three main steps (Figure 7.1) (Fetter, 1994). First, the surface drainage is modified so that runoff moves around the source. If it is not diverted, runoff can erode the soil cover and transport waste constituents. Second, a low permeability cap must be placed over the source. Usually, the cap consists of compacted clay. The purpose of the cap is to prevent additional leachate from forming.

The third step to controlling the source is to create a barrier around the waste body to confine it laterally. Lateral containment can be achieved with slurry walls, grout curtains, or interlocking sheet piling. Alternatively, pumping wells can lower the water table beneath the source to prevent interaction of the waste with groundwater. A slurry wall is constructed by digging a trench around the source and backfilling it with a mixture of water and bentonitic clay. The low permeability medium inhibits the movement of groundwater into the waste body. The depth of a slurry wall is limited by the capability of the digging equipment. Modern devices, resembling a large chain saw, can reach depths beyond 20 m.

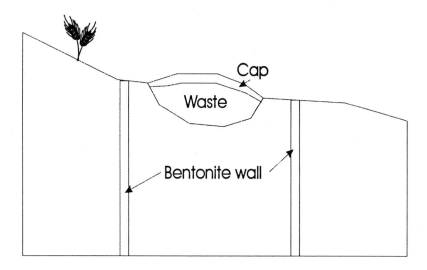

Figure 7.1. Isolated source.

A grout curtain is established by injecting a bentonitic slurry into a series of boreholes. The slurry seeps into the pore spaces of the formation. The technique is not effective for geologic units with tight pores or low effective porosity. Slurry in grout curtains can be injected to great depths below the land surface. It can also be pumped into boreholes that have been drilled through the waste body to form a bottom seal. The main drawback of grout curtains is that slurry in adjacent boreholes may not merge, leaving a gap between holes. Consequently, trenching devices are usually more effective in building slurry barriers.

Alternatively, metal sheet piling can be driven into the soil around the perimeter of a waste body. Interlocking sheets can form an effective cut-off wall, but the method is not appropriate for solid bedrock. The depth of penetration is also limited by the capability of the driving device.

Once the source of contamination is controlled, the contaminated groundwater can be treated. One option is to let contaminants flush through the aquifer by natural recharge and dilution. However, this passive approach is inappropriate if the aquifer is used for drinking water, or if it could become a source of drinking water in the future. It is also inappropriate if the aquifer discharges to surface water. The time for natural restoration could be tens to hundreds of years. A better option is to actively remediate the problem by using in-situ or pumping techniques.

In-situ methods involve treating the contaminant in place. This can be accomplished by using chemical or biological methods. Nutrients and oxygen can be injected into the plume to promote bioremediation (Figure 7.2). Natural soil bacteria with the proper mix of nutrients can biologically treat some compounds such as hydrocarbons.

In cases where the water is shallow, a permeable treatment bed can be installed in the path of a plume (Figure 7.3). For example, an acidic leachate could be neutralized by a permeable bed of limestone gravel. Materials with ion exchange capabilities could be used to remove heavy metals such as lead or cadmium.

Pumping techniques often employ wells to bring contaminated water to the land surface. At the land surface, contaminated water can be treated by various methods. Synthetic organic compounds are often removed by volatilization to the atmosphere in an air-stripping column, adsorption onto activated carbon, or biological treatment.

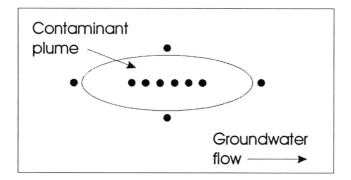

Figure 7.2. Bioremediation scheme (top view); dots represent wells injecting oxygen-enriched water. Interior wells target highest concentrations; peripheral wells control offsite migration.

Typically, the pumping rates and spacing between wells in a groundwater extraction system are determined by computer modeling. In a plume capture scheme, the objectives are to extract the plume, prevent offsite migration, and remove little uncontaminated groundwater.

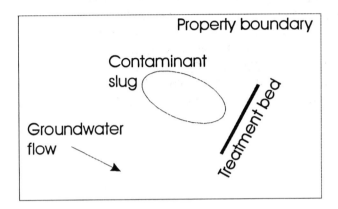

Figure 7.3. Permeable treatment bed (top view).

Previous studies using computer simulation models have shown that a configuration of interior injection wells (injecting oxygen-enriched water) outperforms many alternative schemes in remediating hydrocarbon-contaminated groundwater (Hudak and White, 1997). Moreover, interior injection schemes are effective for a variety of aquifer materials. An alternative pattern of interior extraction wells often creates stagnant zones of low velocity. Contaminants get tied up in these zones between pumping wells and stay there for long periods of time. It is also advantageous to inject water enriched with oxygen to enhance biodegradation.

Figures 7.4 through 7.6 illustrate alternative methods for removing a conservative solute plume (Hudak, 1997b). A computer model was used to simulate removal of the plume in Figure 7.4. A configuration of seven extraction wells, distributed throughout the plume, required 15 years to remove the solute (Figure 7.5). This scheme performed poorly because neighboring wells competed for contaminant, creating stagnant zones of low velocity in which contaminants lingered. An alternative configuration, employing only one extraction well and two injection wells, removed the plume in only three years (Figure 7.6). Yet the total pumping rate was identical for both alternatives. This illustrates the importance of designing proper well layouts for cleaning up contaminated groundwater.

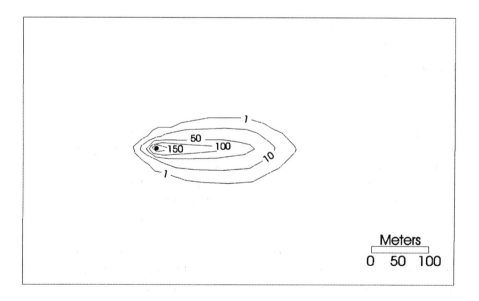

Figure 7.4. Initial contaminant plume; concentration contours in mg/L (Hudak, 1997b – Figure 2, page 22). Reprinted with permission from Springer-Verlag.

Although ineffective for cleaning many sites, extraction wells can reduce contaminant mass while controlling off-site migration. Hydraulic stabilization schemes prevent further migration of contaminants by forcing the hydraulic gradient toward the internal part of the plume. However, extraction wells are generally ineffective for slow-moving groundwater, or when contaminants adsorb to aquifer solids (API, 1992).

Relatively new aquifer remediation strategies include air sparging and steam flushing, in which air or steam is injected into the saturated zone to vaporize volatile compounds. Chemicals spilled from underground or aboveground tanks often contain volatile compounds.

In recent years, it has been recognized that cleaning aquifers to health-based standards is not practical or even feasible at many sites (NRC, 1994). Technical reasons for this difficulty include heterogeneous conditions, NAPLs, contaminants migrating to inaccessible regions, adsorption, and inadequate knowledge of subsurface conditions (NRC, 1994). Heterogeneous conditions make it difficult to predict migration pathways. NAPLs tend to form immobile globules and may migrate deeply underground. They are very difficult to target. Diffusion allows contaminants to move to areas inaccessible to flowing groundwater.

Increasingly, hydrogeologists are using risk-based analysis to determine the level of effort warranted at contaminated sites. Risk-based corrective action is a consistent decision-making process for assessing and responding to a contaminant release, based on the protection of human health and the environment (ASTM, 1995). Within that framework, corrective action activities are tailored to site-specific conditions and risks. The decision process integrates risk and exposure assessment practices with site assessment activities and the selection of remedial measures. Ultimately, the chosen action(s) should optimally utilize limited resources while protecting humans and the environment.

Risk-based programs focus limited resources on those sites posing the greatest threat to human health and the environment. However, remedial actions must achieve an acceptable degree of exposure and risk reduction at all sites.

A risk-based analysis entails several tasks performed by hydrogeologists. For example, they survey sites and surrounding areas for sources of contamination, potential migration pathways, and environmental receptors (that could be impacted by contaminants). Hydrogeologists also must evaluate the proximity of a site to water supplies, collect and interpret soil and water samples, measure aquifer properties that affect the movement of groundwater and contaminants, conduct modeling studies to predict the fate and transport of contaminants, and evaluate potential impacts to environmental receptors.

Based upon data collected and interpreted during a risk analysis, hydrogeologists identify an appropriate response action. Table 1 lists examples of initial response actions for petroleum release sites. The initial response actions are tailored to threats posed by a particular scenario.

Many sites have been closed with gasoline still present in groundwater or soil because (1) it was not feasible to clean the site within a reasonable amount time, (2) it would be too costly to clean the site, and (3) the site posed minimal health risks. Studies in several states show that natural processes limit the extent of hydrocarbon migration in groundwater.

In a comprehensive review of 19,000 leaking tanks in Texas, Mace and others (1997) found that 75 percent of the benzene plumes (benzene is a soluble constituent of gasoline) were less than 80 m long, and only 3% were increasing in length. However, gasoline plumes that reach gravel-filled, buried utility lines or stormwater drains can travel hundreds of meters from a site.

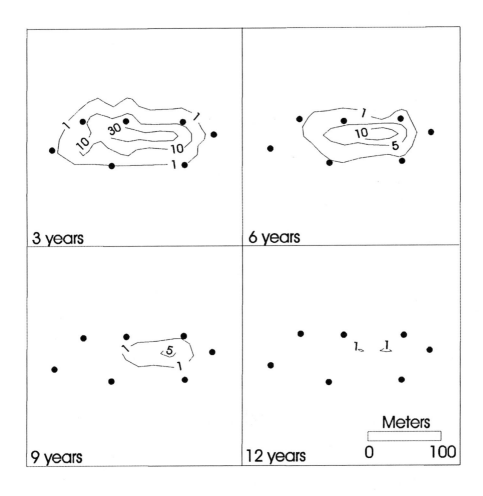

Figure 7.5. Plume reduction due to extraction system; concentration contours in mg/L (Hudak, 1997b – Figure 5, page 23). Reprinted with permission from Springer-Verlag.

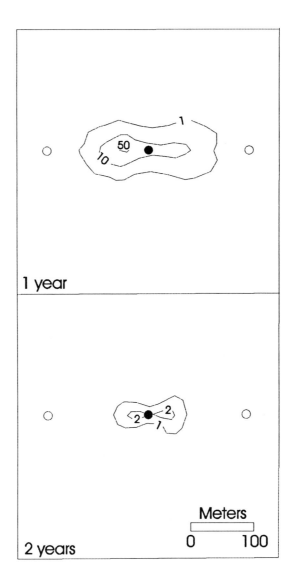

Figure 7.6. Plume reduction due to injection/extraction system; open circles represent injection wells; concentration contours in mg/L (Hudak, 1997b – Figure 6, page 23). Reprinted with permission from Springer-Verlag.

Table 7.1. Example site classification and initial response actions (modified from Johnson and others, 1993; ASTM, 1995).

Criteria/Scenarios	Action
1. Immediate threat to human health, safety, or sensitive environmental receptors	Notify appropriate authorities, property owners, and potentially affected parties.
Explosive levels, or concentrations of vapors that could cause acute health effects, are present in a residence or building.	Evacuate occupants and begin abatement measures such as subsurface ventilation or building pressurization.
Explosive levels of vapors are present in subsurface utility system(s), but no building or residences are impacted.	Evacuate immediate vicinity and begin abatement measures such as ventilation.
Free-product is present in significant quantities at ground surface, on surface water bodies, in utilities other than water supply lines, or in surface water runoff.	Prevent further free-product migration by appropriate containment measures, institute free-product recovery, and restrict area access.
An active public water supply well, public water supply line, or public surface water intake is impacted or immediately threatened.	Notify user(s), provide alternate water supply, hydraulically control contaminated water, and treat water at point-of-use.

Table 7.1. (continued)

Criteria/Scenarios	Action
Ambient vapor/particulate concentrations exceed concentrations of concern from an acute exposure or safety viewpoint.	Install vapor barrier (capping, foams, and so forth), remove source, or restrict access to affected area.
A sensitive habitat or sensitive resources (sport fish, economically important species, threatened and endangered species, and so forth) are impacted and affected.	Minimize extent of impact by containment measures and implement habitat management to minimize exposure.
2. Short-term (0 to 2 years) threat to human health, safety, or sensitive environmental receptors	Notify appropriate authorities, property owners, and potentially affected parties.
There is potential for explosive levels, or concentrations of vapors that could cause acute effects, to accumulate in a residence or other building.	Assess the potential for vapor migration (through modeling) and remove source (if necessary), or install migration barrier.
Shallow contaminated surface soils are open to public access, and dwellings, parks, playgrounds, day-care centers, schools, or similar use facilities that are within 150 m of those soils.	Remove soils, cover soils, or restrict access.

Table 7.1. (continued)

Criteria/Scenarios	Action
A non-potable water supply well is impacted or immediately threatened.	Notify owner/user and evaluate the need to install point-of-use water treatment, hydraulic control, or alternate water supply.
Groundwater is impacted, and a public or domestic water supply well producing from the impacted aquifer is located within two-years projected groundwater travel distance downgradient of the known extent of chemical(s) concern.	Institute monitoring and then evaluate if natural attenuation is sufficient, or if hydraulic control is required.
Groundwater is impacted, and a public or domestic water supply well producing from a different interval is located within the known extent of chemicals of concern.	Monitor groundwater well quality and evaluate if control is necessary to prevent vertical migration to the supply well.
Impacted surface water, storm-water, or groundwater discharges within 150 m of a sensitive habitat or surface water body used for human drinking water or contact recreation.	Institute containment measures, restrict access to areas near discharge, and evaluate the magnitude and impact of the discharge.

Table 7.1. (continued)

Criteria/Scenarios	Action
3. Long-term (>2 years) threat to human health, safety, or sensitive environmental receptors	Notify appropriate authorities, property owners, and potentially affected parties.
Subsurface soils >1 m beneath ground surface are significantly impacted, and the depth between impacted soils and the first potable aquifer is <15 m.	Monitor groundwater and determine the potential for future migration of the chemical(s) to the aquifer.
Groundwater is impacted, and potable water supply wells producing from the impacted interval are located >2 years groundwater travel time from the dissolved plume.	Monitor the dissolved plume and evaluate the potential for natural attenuation and the need for hydraulic control.
Groundwater is impacted, and non-potable water supply wells producing from the impacted interval are located >2 years groundwater travel time from the dissolved plume.	Identify use of well, assess the effect of potential impact, monitor the dissolved plume, and evaluate whether natural attenuation or hydraulic control are appropriate control measures.
Groundwater is impacted, and non-potable water supply wells not producing from the impacted interval are within the known extent of chemicals.	Monitor the dissolved plume, determine the potential for vertical migration, notify the user(s), and determine if any impact is likely.

Table 7.1. (continued)

Criteria/Scenarios	Action
Impacted surface water, storm-water, or groundwater discharges within 450 m of a sensitive habitat or surface water body used for human drinking water or recreation.	Investigate current impact on sensitive habitat or surface water body, restrict access to area of discharge (if necessary), and evaluate the need for containment/control measures.
Shallow contaminated surface soils are open to public access, and dwellings, parks, playgrounds, day-care centers, schools, or similar use facilities that are >150 m from those soils.	Restrict access to impacted soils.
4. No demonstrable long-term threat to human health or safety or sensitive environmental receptors.	Notify appropriate authorities, property owners, and potentially affected parties.
Non-potable aquifer with no existing local use impacted.	Monitor groundwater and evaluate effect of natural attenuation on dissolved plume migration.

Table 7.1. (continued)

Criteria/Scenarios	Action
Impacted soils located more than 1 m below ground surface and more than 15 m above nearest aquifer.	Monitor groundwater and evaluate effect of natural attenuation on leachate migration.
Groundwater is impacted, and non-potable wells are located downgradient of the known extent of the chemical(s) of concern, and they produce from a non-impacted zone.	Monitor groundwater and evaluate effect of natural attenuation on dissolved plume migration.

Geographic information systems are useful for keeping track of soil, construction, and monitoring data that affect underground storage tank integrity (Hudak and others, 1995). These computer-based systems can be used by planning and regulatory agencies to determine which sites may pose the greatest environmental hazard and warrant immediate attention. For example, Figure 7.7 shows several steel, underground tanks in Denton, Texas that (at the time the map was created) were over 20 years old. Using a geographic information system, the tank map was laid over a map of land resource units (including a sandstone aquifer and various aquitards). Recently, an EPA deadline passed by which tank owners had to upgrade or properly abandon their tanks. Upgrades included spill guards, corrosion protection, containment, and monitoring devices to detect petroleum releases.

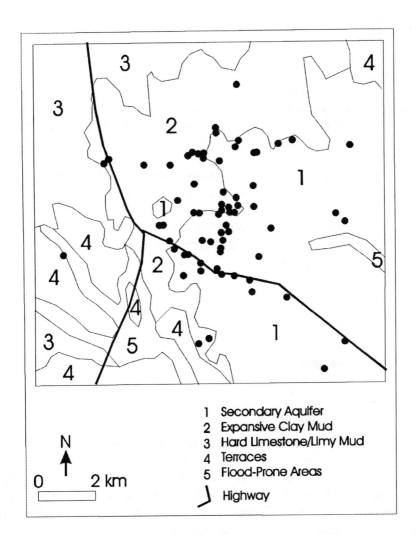

Figure 7.7. Underground tanks more than 20 years old in relation to land resource units, Denton, TX (Hudak and others, 1995). Reprinted with permission from American Water Resources Association.

Capture Zones

Capture zones delineate the horizontal area of an aquifer that contributes water to a pumping well. They are useful for designing aquifer remediation schemes, and also for defining protection areas around municipal water supply wells. Cities that rely on groundwater may restrict land uses within a supply well's capture zone to reduce the chance of contamination.

An infinite-time capture zone delineates a domain within which groundwater eventually reaches a pumping well. It has (approximately) a parabolic shape that is open and elongate in the upgradient direction (Figure 7.8). Todd (1980) gave an equation for calculating the shape of an infinite-time capture zone in a confined aquifer with steady flow:

$$x = \frac{-y}{\tan(2\pi K b i y / Q)} \qquad (7.1)$$

In Equation 7.1, x and y are Cartesian coordinates on the capture zone boundary. The origin of the coordinate system coincides with the pumping well. The regional hydraulic gradient prior to pumping is denoted by i, the saturated thickness by b, and other variables retain previous definitions. Radians rather than degrees apply to the quantity after the tangent function.

The distance from the pumping well to the downgradient edge of the capture zone is

$$x_0 = -Q / (2\pi K b i) \qquad (7.2)$$

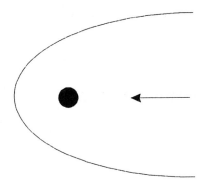

Figure 7.8. Infinite-time capture zone.

In the upgradient direction, the capture zone attains a maximum half-width of

$$y_{max} = \pm Q /(2Kbi) \tag{7.3}$$

Example 7.1: Compute the downgradient edge and maximum half-width of a capture zone in a confined aquifer, given the following information:

b = 54 m
i = 0.001
Q = 245 m³/d
K = 0.7 m/d

$$x_0 = -Q/(2\pi Kbi) = \frac{-245 \text{ m}^3}{d} \times \frac{1}{2} \times \frac{1}{\pi} \times \frac{1 d}{0.7 \text{ m}} \times \frac{1}{54 \text{ m}} \times \frac{1}{0.001} =$$

$$-1,032 \text{ m}$$

$$y_{max} = \pm Q/(2Kbi) = \pm \frac{245 \text{ m}^3}{d} \times \frac{1}{2} \times \frac{1 d}{0.7 \text{ m}} \times \frac{1}{54 \text{ m}} \times \frac{1}{0.001} =$$

$$\pm 3,241 \text{ m}$$

For an unconfined aquifer, an infinite-time capture zone can be obtained from Equation 7.4 (Grubb, 1993):

$$x = \frac{-y}{\tan[\pi K(H_1^2 - H_2^2)y/(QM)]} \qquad (7.4)$$

where H_1 and H_2 are hydraulic head values, referenced to the base of the aquifer, along a flow path before pumping. H_1 is directly up-gradient and H_2 downgradient of the well. M is the horizontal distance between the points at which H_1 and H_2 are measured.

The maximum half-width of the capture zone as x approaches infinity is

$$y_{max} = \pm \frac{QM}{K(H_1^2 - H_2^2)} \qquad (7.5)$$

Finally, the distance from the pumping well to the downgradient edge of the capture zone is

$$x_0 = \frac{-QM}{\pi K (H_1^2 - H_2^2)} \tag{7.6}$$

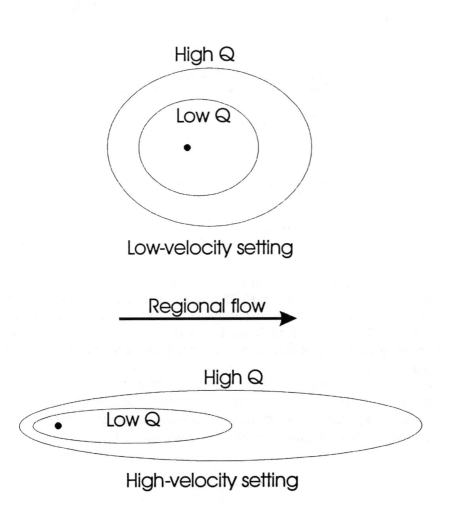

Figure 7.9. Effect of pumping rate (Q) and groundwater velocity on capture zone geometry. (Single-well capture zones in uniform flow fields often resemble ellipses, but are not precisely elliptical.)

Example 7.2: Calculate the downgradient edge and maximum half-width of a capture zone in an unconfined aquifer. The aquifer has a hydraulic conductivity of 0.08 m/day, $M = 78$ m, $H_1 = 23$ m, and $H_2 = 17$ m. The pumping rate is 14 m³/day.

$$x_0 = \frac{-QM}{\pi K(H_1^2 - H_2^2)} = \frac{-14\,\text{m}^3}{\text{day}} \times 78\,\text{m} \times \frac{1}{\pi} \times \frac{1\,\text{d}}{0.08\,\text{m}} \times$$

$$\frac{1}{[(23\,\text{m})^2 - (17\,\text{m})^2]} = -18.1\,\text{m}$$

$$y_{max} = \pm\frac{QM}{K(H_1^2 - H_2^2)} = \pm\frac{14\,\text{m}^3}{\text{day}} \times 78\,\text{m} \times \frac{1\,\text{d}}{0.08\,\text{m}} \times$$

$$\frac{1}{[(23\,\text{m})^2 - (17\,\text{m})^2]} = \pm56.9\,\text{m}$$

Time-dependent capture zones are not open-ended, but closed at the upgradient end. The shape of a time-dependent capture zone depends on the groundwater velocity and pumping rate (Figure 7.9). While time-dependent capture zones are useful for designing pumping schemes, simply superimposing several capture zones having an identical shape over a map of a contaminant plume is inappropriate. This procedure incorrectly portrays the effects of multiple pumping wells, which yield complex capture zones. Stagnation zones that develop between neighboring extraction wells delay contaminant removal (Hudak, 1997b).

Key Terms

air sparging, capture zone, corrective action plan, extraction wells, geographic information systems, grout curtain, risk-based analysis, sheet piling, in-situ remediation, slurry wall, source control, steam flushing

Problems

1. Compute the infinite-time capture zone of an extraction well pumping 220,000 gal/day from a confined aquifer with a hydraulic conductivity of 1,000 ft/d, hydraulic gradient of 0.002, and saturated thickness of 60 ft. Sketch the capture zone on a sheet of graph paper. Note the position of the well.

2. Describe the steps involved in cleaning a contaminated aquifer.

3. Describe an ideal situation for excavating and removing a contaminant source in lieu of alternative source control options.

4. Under what conditions would passive remediation be an appropriate groundwater management strategy?

5. What types of groundwater contamination can be biologically treated?

6. Compute an infinite-time capture zone for an unconfined aquifer given the following data.

    Q = 40,000 ft³/d
    K = 10 ft/d
    M = 270 ft
    H_1 = 120 ft
    H_2 = 101 ft

7.   Suppose that a military base has contaminated groundwater
      in an underlying aquifer. Domestic water supply wells pro-
      ducing from the impacted aquifer are more than two years
      groundwater travel time from the contaminant plume. Ac-
      cording to Table 7.1, what might be an appropriate initial re-
      sponse action. What other information would you need to
      devise an appropriate plan of action? What are some of the
      potential problems with using a risk-based approach to
      groundwater cleanup. Cite some advantages and disadvan-
      tages of a "brute force" approach, in which a site is actively
      remediated until their are no traces of contamination.

CHAPTER 8

Careers in Hydrogeology

Hydrogeologists are employed by various public and private sector organizations. Private consulting firms are the largest source of employment for groundwater specialists. These companies are often involved in characterizing or remediating contaminated soil and groundwater. Typical job activities include drilling and logging holes, installing monitoring wells, measuring water levels and collecting samples, conducting slug and pump tests, analyzing data in the office, and writing reports. In the office, hydrogeologists routinely use computers to aid in mapping sites, keeping track of data, and modeling contaminant fate and transport. Modeling is important for assessing the risk a site poses to humans and other environmental receptors. Such modeling is also used to investigate potential remediation schemes.

Consultants often interface with local, state, and federal environmental regulatory agencies. Examples of such agencies include the Environmental Protection Agency, Pennsylvania Department of Natural Resources, and Texas Natural Resource Conservation Commission. These agencies are an additional source of employment for groundwater scientists. Individuals working for regulatory agencies do a combination of office and field work. Office work involves keeping track of files and reporting deadlines for numerous hazardous waste sites. Field work includes visiting sites to ensure they are in compliance with current laws.

Industries that generate hazardous waste also employ hydrogeologists. Typically, these specialists help the companies make appropriate decisions on storing and transporting waste. Their objective is to protect the environment, comply with the relevant legal framework, and save the company money over the long term. A large contamination problem could cost a company hundreds of thousands of dollars. Indeed, many companies have declared bankruptcy in the face of impending soil and groundwater clean-up costs.

Hydrogeologists who acquire doctoral degrees may find employment in the academic sector. Usually, this entails teaching and research at a four-year college or university. Typical job functions include developing and teaching classes, writing and publishing articles, making presentations at professional meetings, and community outreach. Many doctoral scientists who don't work at universities are involved in basic and applied research for government agencies such as the Department of Energy and U.S. Geological Survey. These researchers develop innovative methods for cleaning subsurface contamination, and compile and evaluate new data on aquifers.

Among the highest paying groundwater-related occupations is environmental law. Attorneys who have obtained a law degree and a graduate degree in groundwater litigate cases involving environmental contamination that may have adversely impacted humans or other organisms.

Finally, water supply agencies and planning organizations often employ groundwater specialists to manage water resources. These specialists may determine appropriate locations for new wells, or recommend locations for waste repositories that pose a threat to groundwater. Often, their goal is maximize aquifer yield while avoiding overdrafting or water quality degradation.

Key Terms

consulting firms, environmental law, environmental regulatory agencies, water resources

Problems

1.   Check a local phone book or the Internet, and compile a list of 10 agencies that employ hydrogeologists.

2.   Contact one of the companies on your list, and write down a few of the services they provide.

SYMBOLS

Symbol	Definition	Units
b	saturated thickness	length
g	acceleration of gravity	length/time²
h	unrecovered head differential, slug test	length
h_e	elevation head	length
h_f	final water level above outlet, falling-head permeameter test	length
h_i	initial water level above outlet, falling-head permeameter test	length
h_p	pressure head	length
i	hydraulic gradient	none
k	intrinsic permeability	length²
n	total porosity	none
n_e	effective porosity	none
r	distance between pumping well and observation well, transient pumping test in confined aquifer	length
r_i	radius of intake, slug test	length
r_p	radius of piezometer, slug test	length
s	drawdown	length
Δs	drawdown over one log cycle	length
t	time	time
v	groundwater velocity	length/time
x	Cartesian coordinate	length
x_0	distance from pumping well to stagnation point	length
y	Cartesian coordinate	length
y_{max}	maximum half-width of capture zone	length
z	vertical distance from sea level	

	to freshwater/saltwater inter- face	length
A	area of aquifer	length^2
A_c	cross-sectional area of falling- head chamber	length^2
A_t	cross-sectional area of falling- head tube	length^2
B	conversion factor	none
C	effluent concentration	mass/length^3
C_0	influent concentration	mass/length^3
C_m	conductance in microsiemens	
$C_{solution}$	concentration in solution, batch experiment	mass/length^3
C_{soil}	concentration in soil, batch experiment	mass/length^3
D_L	coefficient of longitudinal hydrodynamic dispersion	length^2/time
D_{TH}	coefficient of transverse- horizontal hydrodynamic dispersion	length^2/time
D_{TV}	coefficient of transverse- vertical hydrodynamic dispersion	length^2/time
D^*	effective molecular diffusion coefficient	length^2/time
G	specific gravity of gasoline	none
H	hydraulic head	length
H_0	initial head differential, slug test	length
H_1	hydraulic head at closer observation well	length
H_2	hydraulic head at further observation well	length
I	inputs to hydrologic system	length^3
K	hydraulic conductivity	length/time
K_1	hydraulic conductivity of layer 1	length/time
K_2	hydraulic conductivity of	

	layer 2	length/time
K_d	distribution coefficient	length^3/mass
L	length of sample	length
L_i	length of intake, slug test	length
M	horizontal distance between observation wells in unconfined aquifer	length
O	outputs from hydrologic system	length^3
Q	discharge	length^3/time
R_1	distance to closer observation well	length
R_2	distance to further observation well	length
R_f	retardation factor	none
S	storage coefficient	none
ΔS	change in storage in hydrologic system	length^3
S_c	specific capacity	length^2/time
S_r	specific retention	none
S_y	specific yield	none
T	transmissivity	length^2/time
T_0	time at which $h/H_0 = 0.37$ on semi-log plot	time
V	total volume of aquifer sample	length^3
V_g	volume of water drained by gravity	length^3
V_r	volume of water retained against gravity	length^3
$W(u)$	well function	none
α_L	longitudinal dispersivity	length
α_{TH}	transverse-horizontal dispersivity	length
α_{TH}	transverse-vertical dispersivity	length
ρ	density of water	mass/length^3
ρ_b	dry bulk density	mass/length^3
ρ_g	density of gasoline	mass/length^3

ρ_s	particle density	mass/length^3
ρ_t	density of saltwater	mass/length^3
θ_1	angle between flowline and perpendicular to interface, layer 1	degrees
θ_2	angle between flowline and perpendicular to interface, layer 2	degrees
μ	viscosity of water	mass/time/ length

GLOSSARY

adsorption Adhesion of solutes from a solution to aquifer solids.

advection Movement of solutes with groundwater.

air rotary A drilling method employing a high-speed rotating bit and air to circulate cuttings.

air sparging Pumping air into an aquifer to volatilize contaminants.

ambient monitoring Monitoring existing conditions in an aquifer.

anisotropic The magnitude of an aquifer property varies with direction about a point.

aquifer A formation producing useful quantities of water.

arbitrary boundary A flow net boundary not coinciding with a physical feature of an aquifer.

bucket auger A manual drilling device with a hollow head and cutting blades.

bulk density Dry mass divided by the total volume of a soil sample.

cable tool A drilling device employing a heavy drill string and chisel-like bit repeatedly raised and lowered into the ground.

capillary fringe A zone immediately above the water table in which water is drawn upward under negative fluid pressure.

capture zone The part of an aquifer that contributes water to a pumping well.

case-preparation monitoring Monitoring groundwater to obtain data for a lawsuit.

casing A pipe used to construct a well.

chain of custody A form that shows who collected and received groundwater samples.

concentration A measure of the amount of solute in water.

cone of depression The curved water table or potentiometric surface that develops around a pumping well.

confined aquifer An aquifer overlain by a confining layer.

confining layer A low-permeability layer that bounds an aquifer.

conservative solute A solute that does not adsorb onto aquifer solids or undergo chemical reactions.

constant-head boundary An aquifer boundary along which the hydraulic head is constant.

constant-head permeameter A device used to measure the hydraulic conductivity of an aquifer sample, under constant-head conditions.

consulting firms Private companies that employ hydrogeologists to solve groundwater problems.

corrective action plan A plan devised to clean a contaminated site.

Darcy's law The volumetric discharge of groundwater is equal to the product of hydraulic conductivity, hydraulic gradient, and cross-sectional area of aquifer perpendicular to flow.

diffusion Movement of solutes in groundwater from areas of high concentration to low concentration.

direct approach An approach to aquifer testing in which estimated aquifer properties are used to predict drawdown.

discharge area An area from which groundwater is leaving an aquifer.

dispersivity A measure of the ability of an aquifer to spread solutes by mechanical means.

distribution coefficient A measure of the tendency of a solute to attach to aquifer solids.

DNAPL Dense, nonaqueous phase liquid.

drawdown The amount the hydraulic head drops in an aquifer.

driven well A metal pipe with a conical tip driven into the ground with a hammering device.

dual-wall reverse-circulation A drilling method that circulates fluids down the annulus of a hole and up the drill stem.

effective porosity The volume of interconnected pore space in an aquifer sample divided by the total volume of the sample.

effluent river A river that receives groundwater.

electrical conductance The ability of rock, soil, or water to conduct an electrical current.

elevation head The elevation of a piezometer where it is open to an aquifer.

environmental law Laws pertaining to the environment.

environmental regulatory agencies Federal, state, or local agencies that enforce environmental laws.

equipotential line A line along which the hydraulic head is constant.

evaporation Transfer of water from the liquid phase to the vapor phase.

extraction wells Wells used to extract water from aquifers.

falling-head permeameter A device used to measure the hydraulic conductivity of an aquifer sample under falling-head conditions.

field blank A vial of purified water taken to the field and run through the sampling equipment.

filter pack Coarse sand packed around the screen of a well.

flow line A line depicting the path of flowing groundwater.

flow net A two-dimensional illustration consisting of equipotential lines and groundwater flow lines.

geographic information system A computer-based software package capable of storing, displaying, and querying location and attribute information for points, lines, or polygons.

Ghyben-Herzberg principle The distance from sea level to the freshwater/saltwater interface is approximately 40 times the distance from sea level to the water table.

groundwater Water present in the saturated zone beneath Earth's surface.

groundwater velocity The rate at which groundwater travels.

grout curtain A series of holes filled with bentonite used to create a groundwater flow barrier.

heterogeneous The magnitude of an aquifer property is spatially variable.

hollow-stem A drilling method employing hollow augers.

homogeneous The magnitude of an aquifer property is spatially uniform.

hydraulic conductivity The product of intrinsic permeability, fluid density, and the acceleration of gravity, divided by fluid viscosity.

hydraulic gradient The difference in the hydraulic head between two points along a flow line, divided by the distance between the points.

hydraulic head The elevation of water in a piezometer, pertaining to the point at which the piezometer is open to an aquifer.

hydrodynamic dispersion Spreading of groundwater and its dissolved constituents by mechanical mixing and chemical diffusion.

hydrogeology The study of groundwater.

hydrologic budget Volumetric inputs minus outputs from a hydrologic system are equal to the change in storage in the system.

hydrologic cycle Continuous circulation of Earth's water.

infiltration Movement of water from Earth's surface into the vadose zone.

influent river A river that loses water to the subsurface.

in-situ remediation Cleaning contaminants in place.

intrinsic permeability A measure of the ability of an aquifer to transmit fluids.

inverse approach An approach to aquifer testing in which drawdown observations are used to estimate aquifer properties.

isotropic The magnitude of an aquifer property is uniform in all directions about a point.

jet percussion A drilling method employing a high-speed water stream.

leachate Chemicals dissolved from solid waste.

LNAPL Light, nonaqueous phase liquid.

maximum contaminant level The concentration above which a contaminant poses a health concern.

monitoring well A cylindrical pipe open at the bottom used to measure water levels and collect groundwater samples.

multilevel sampling A sampling device open to an aquifer at different elevations.

nitrate A common groundwater contaminant, often derived from decaying vegetation, fertilizer, and animal waste.

no-flow boundary A boundary that cannot be traversed by flowing groundwater.

particle density The density of solid particles.

piezometer A pipe inserted into the saturated zone to measure the hydraulic head.

potentiometric surface The level to which water rises in piezometers tapping a confined aquifer.

potentiometric surface map An illustration in map view, consisting of equipotential lines and flow lines, depicting the movement of groundwater at some depth interval.

precipitation Falling products of condensation in Earth's atmosphere.

pressure head The height of the water column in a piezometer, pertaining to the point at which the piezometer is open to an aquifer.

primary openings Pores between solid particles in a rock or unconsolidated deposit that were present when the rock or deposit formed.

pumping test A test in which a well tapping an aquifer is pumped, and drawdown is measured in observation wells tapping the same aquifer.

pumping well A well from which groundwater is pumped.

reactive solute A solute that tends to adsorb to aquifer solids or react chemically.

recharge area An area in which groundwater enters an aquifer.

refraction Bending of groundwater flow lines.

research monitoring Monitoring to learn about the basic behavior of groundwater and the chemicals it transports.

retardation factor A dimensionless number used to express the relative velocity of a chemical in groundwater.

risk-based analysis Evaluating the human and environmental risk of a contaminated site to determine an appropriate course of action.

runoff Water flowing over Earth's surface.

saltwater intrusion Underground saline water seeping into a freshwater well.

saturated zone The interval beneath the vadose zone, in which all of the openings are filled with water.

screen The slotted, open part of a well.

secondary openings Cracks or cavities in a rock or unconsolidated deposit that developed after the rock or deposit formed.

seepage meter A device used to estimate the rate at which water seeps into or out of the bottom or sides of a surface water body.

sheet flow Water flowing in a thin film over Earth's surface.

sheet piling Metal sheets driven into the ground to block groundwater flow.

Shelby tube A thin, cylindrical tube driven into the ground to extract a soil sample.

slug test A test in which the water column is displaced in a well, and then repeatedly measured at it migrates back to the original position.

slurry wall A trench filled with bentonite clay to block moving groundwater.

solid-stem A drilling method employing solid augers.

solute Anything dissolved in water.

source control Removing, enclosing, or controlling a source of contamination to prevent further subsurface pollution.

source monitoring Monitoring groundwater near a potential source of contamination.

specific capacity The discharge of a well divided by the stabilized drawdown in the well.

specific retention The volume of water a saturated sample retains against gravity, divided by the total volume of the sample.

specific yield The volume of water a saturated sample yields to gravity, divided by the total volume of the sample.

spiked sample A sample containing a known concentration of a particular solute.

split sample A sample that is split into two smaller samples.

split-spoon A soil sampling device consisting of a cylinder that breaks lengthwise into two halves.

springs Places where groundwater seeps out at the land surface.

steady flow Groundwater flow in which hydraulic head does not change over time.

steam flushing Pumping steam into a contaminated aquifer to remove organic compounds.

storage coefficient The volume of water an aquifer sample releases, per unit surface area, per unit decline in hydraulic head.

stream discharge The volume of water flowing past a stream's cross-section, per unit time.

sublimation The transfer of water from the solid phase to the vapor phase.

TDS Total dissolved solids.

tensiometer A device used to measure fluid pressure in the vadose zone.

total porosity The volume of pores in an aquifer sample, divided by the total volume of the sample.

transient flow Groundwater flow in which the hydraulic head changes over time.

transmissivity The product of hydraulic conductivity and saturated thickness.

transpiration Release of water vapor from plants to the atmosphere.

tremie pipe A narrow pipe used to place annular material around a well casing.

trend test Measuring temporal changes in groundwater levels before conducting an aquifer test.

unconfined aquifer An aquifer not overlain by a confining layer.

vadose zone The zone immediately underlying Earth's surface, in which not all of the openings are filled with water.

vadose zone monitoring Deploying monitoring devices in the vadose zone.

variable-head boundary An aquifer boundary along which the hydraulic head attains different values.

viscosity A measure of the ability of a fluid to resist flow.

water and mud rotary Drilling methods that employ a high-speed rotating bit and water or mud to circulate cuttings.

water resources Water used by humans or other organisms.

water table An underground surface at which (gauge) fluid pressure is equal to zero.

water table map An illustration in map view, consisting of water table contours and flow lines, depicting the movement of groundwater in the upper interval of an unconfined aquifer.

well evacuation Removing stagnant water from a well before sampling.

REFERENCES

API (American Petroleum Institute), 1992. *Pump and Treat: The Petroleum Industry Perspective*. American Petroleum Institute, Washington, DC.

ASTM (American Society for Testing and Materials), 1995. *Standard Guide for Risk-Based Corrective Action Applied at Petroleum Release Sites*. American Society for Testing and Materials, West Conshohocken, PA.

Barcelona, J. J., Gibb, J. P., and Miller, R. A., 1984. A guide to the selection of materials for monitoring well construction and groundwater sampling. *Illinois State Water Survey Contract Report*, 327.

Bouwer, H., 1978. *Groundwater Hydrology*. Mc-Graw Hill, New York, NY.

Cullen, S. J., Kramer, J. H., and Luellen, J. R., 1995. A systematic approach to designing a multiphase unsaturated zone monitoring network. *Groundwater Monitoring and Remediation*, Summer Issue, 124-135.

Darcy, H., 1856. *Les Fontaines Publiques de la Ville de Dijon*. Victor Dalmont, Paris.

Davis, S. N., 1969. Porosity and permeability in natural materials. In: *Flow through Porous Media*, DeWiest, R. J. M. (ed.), 53-89. Academic Press, New York, NY.

Driscoll, F. G., 1986. *Groundwater and Wells*. Johnson Division, St. Paul, MN.

EPA (U.S. Environmental Protection Agency), 1991a. *Compendium of ERT Groundwater Sampling Procedures*. U.S. Environmental Protection Agency, Washington, DC.

EPA (U.S. Environmental Protection Agency), 1991b. *Groundwater, Volume 11 - Methodology.* U.S. Environmental Protection Agency, Office of Research and Development, Cincinnati, OH.

EPA (U.S. Environmental Protection Agency), 1994a. *RCRA Groundwater Monitoring - Draft Technical Guidance.* Government Institutes, Rockville, MD.

EPA (U.S. Environmental Protection Agency), 1994b. *National Primary Drinking Water Standards.* Washington, DC.

Fetter, C. W., 1994. *Applied Hydrogeology.* MacMillan, New York, NY.

Freeze, R. A., and Cherry, J. A., 1979. *Groundwater.* Prentice-Hall, Englewood Cliffs, NJ.

Gelhar, L. W., Welty, C., and Rehfeldt, K. R., 1992. A critical review of data on field-scale dispersion in aquifers. *Water Resources Research,* 28(7), 1955-1974.

Grubb, S., 1993. Analytical model for estimation of steady-state capture zones of pumping wells in confined and unconfined aquifers. *Groundwater,* 31(1), 27-32.

Heath, R. C., 1983. Basic groundwater hydrology. *U.S. Geological Survey Water Supply Paper,* 2220.

Hem, J. D., 1970. Study and interpretation of the chemical characteristics of natural water. *U.S. Geological Survey Water Supply Paper,* 1473.

Hudak, P. F., Loaiciga, H. A., and Schoolmaster, F. A., 1993. Application of geographic information systems to groundwater monitoring network design. *Water Resources Bulletin,* 29(3), 383-390.

Hudak, P. F., 1994a. Effective porosity of unconsolidated sand: Estimation and impact on capture zone geometry. *Environmental Geology*, 24, 140-143.

Hudak, P. F., 1994b. A method for monitoring groundwater quality near waste storage facilities. *Environmental Monitoring and Assessment*, 30, 197-210.

Hudak, P. F., 1997a. Geophysical prospecting for oil and gas well effluent. *The Professional Geologist*, 34(13), 2-4.

Hudak, P. F., 1997b. Evaluation of a capture zone overlay method for designing groundwater remediation systems. *Environmental Geology*, 31(1/2), 21-26.

Hudak, P. F., and Loaiciga, H. A., 1991. Mass transport modeling in landfill-contaminated buried valley aquifer. *Journal of Water Resources Planning and Management*, 117(2), 260-272.

Hudak, P. F., Speas, R. K., and Schoolmaster, F. A., 1995. Managing underground storage tanks in urban environments: A geographic information systems approach. *Water Resources Bulletin*, 31(3), 439-445.

Hudak, P. F., and Blanchard, S., 1997. Land use and groundwater quality in the Trinity Group outcrop of north-central Texas. *Environment International*, 23(4), 507-517.

Hudak, P. F., and White, S. A., 1997. Modeling alternative groundwater remediation methods in contrasting hydrogeologic settings. *Journal of Environmental Science and Health*, A32(1), 105-122.

Hudak, P. F., 1998. Procedure for shifting groundwater monitoring locations to enhance detection efficiency near landfills. *Journal of Environmental Science and Health*, A33(8), 1771-1780.

Hvorslev, M. J., 1951. Time lag and soil permeability in ground-water observations. *U.S. Army Corps of Engineers Waterways Experiment Station Bulletin*, 36.

Johnson, A. I., 1967. Specific yield – compilation of specific yields for various materials. *U.S. Geological Survey Water Supply Paper*, 1662-D.

Johnson, D. C., DeVaull, G. E., Ettinger, R. A., MacDonald, R. L. M., Stanley, C. C., Westby, T. X., and Conner, J., 1993. *Risk-Based Corrective Action: Tier 1 Guidance Manual.* Shell Oil Company, Houston, TX.

Lohman, S. W., 1979. Groundwater hydraulics. *U.S. Geological Survey Professional Paper*, 708.

Mace, R. E., Fisher, S. F., Welch, D. M., and Parra, S. P., 1997. *Extent, Mass, and Duration of Hydrocarbon Plumes from Leaking Petroleum Storage Tank Sites in Texas.* Bureau of Economic Geology, Austin, TX.

Nace, R. L., Are we running out of water? *U.S. Geological Survey Circular*, 586.

Neuman, S. P., 1975. Analysis of pumping test data from aniso-tropic unconfined aquifers considering delayed gravity response. *Water Resources Research*, 11, 329-342.

NRC (National Research Council), 1994. *Alternatives for Groundwater Cleanup.* National Academy Press, Washington, DC.

Sanders, L. L., 1998. *A Manual of Field Hydrogeology.* Prentice Hall, Upper Saddle River, NJ.

Solley, W. B., Pierce, R. R., and Perlman, H. A., 1998. Estimated use of water in the United States in 1995. *U.S. Geological Survey Circular*, 1200.

Sun, R. J., 1986. Regional aquifer-system analysis program of the U.S. Geological Survey. *U.S. Geological Survey Circular*, 1002.

Testa, S., and Pacskowski, M., 1989. Volume determination and recoverability of free hydrocarbon. *Groundwater Monitoring and Remediation*, 9(1), 120-128.

Theis, C. V., 1935. The relation between the lowering of the piezometric surface and the rate and duration of discharge of a well using ground-water storage. *Transactions, American Geophysical Union*, 16, 519-524.

Thiem, G., 1906. *Hydrologische Methoden*. Gebhardt, Leipzig.

Todd, D. K., 1980. *Groundwater Hydrology*. John Wiley & Sons, New York, NY.

INDEX